THE CRAFT OF SCIENTIFIC COMMUNICATION

On Writing, Editing, and Publishing
Jacques Barzun

Telling About Society
Howard S. Becker

Tricks of the Trade
Howard S. Becker

Writing for Social Scientists
Howard S. Becker

Permissions, A Survival Guide: Blunt Talk about Art as Intellectual Property
Susan M. Bielstein

The Craft of Translation
John Biguenet and Rainer Schulte, editors

The Craft of Research
Wayne C. Booth, Gregory G. Colomb, and Joseph M. Williams

The Dramatic Writer's Companion
Will Dunne

Glossary of Typesetting Terms
Richard Eckersley, Richard Angstadt, Charles M. Ellerston, Richard Hendel, Naomi B. Pascal, and Anita Walker Scott

Writing Ethnographic Fieldnotes
Robert M. Emerson, Rachel I. Fretz, and Linda L. Shaw

Legal Writing in Plain English
Bryan A. Garner

From Dissertation to Book
William Germano

Getting It Published
William Germano

A Poet's Guide to Poetry
Mary Kinzie

The Chicago Guide to Collaborative Ethnography
Luke Eric Lassiter

Cite Right
Charles Lipson

How to Write a BA Thesis
Charles Lipson

The Chicago Guide to Writing about Multivariate Analysis
Jane E. Miller

The Chicago Guide to Writing about Numbers
Jane E. Miller

Mapping It Out
Mark Monmonier

The Chicago Guide to Communicating Science
Scott L. Montgomery

Indexing Books
Nancy C. Mulvany

Getting into Print
Walter W. Powell

The Subversive Copy Editor
Carol Fisher Saller

A Manual for Writers of Research Papers, Theses, and Dissertations
Kate L. Turabian

Tales of the Field
John Van Maanen

Style
Joseph M. Williams

A Handbook of Biological Illustration
Frances W. Zweifel

The Craft of Scientific Communication

JOSEPH E. HARMON & ALAN G. GROSS

THE UNIVERSITY OF CHICAGO PRESS CHICAGO AND LONDON

JOSEPH E. HARMON is a senior writer and editor at Argonne National Laboratory. He is coeditor of *The Scientific Literature* with Alan G. Gross and coauthor of *Communicating Science* with Gross and Michael Reidy.

ALAN G. GROSS is a professor in the Department of Communication Studies at the University of Minnesota. He has published widely on rhetoric, rhetorical theory, and visual communication. He is author or coauthor of several books, including *The Rhetoric of Science.*

The University of Chicago Press, Chicago 60637
The University of Chicago Press, Ltd., London

Printed in the United States of America

19 18 17 16 15 14 13 12 11 10   1 2 3 4 5

ISBN-13: 978-0-226-31661-1 (cloth)
ISBN-13: 978-0-226-31662-8 (paper)

ISBN-10: 0-226-31661-0 (cloth)
ISBN-10: 0-226-31662-9 (paper)

Library of Congress Cataloging-in-Publication Data

Harmon, Joseph E.
The craft of scientific communication / Joseph E. Harmon and Alan G. Gross.
p. cm.
Includes bibliographical references and index.
ISBN-13: 978-0-226-31661-1 (cloth : alk. paper)
ISBN-13: 978-0-226-31662-8 (pbk. : alk. paper)
ISBN-10: 0-226-31661-0 (cloth : alk. paper)
ISBN-10: 0-226-31662-9 (pbk. : alk. paper) 1. Communication in science—Handbooks, manuals, etc. 2. Science—Language—Handbooks, manuals, etc. 3. Technical writing—Handbooks, manuals, etc. I. Gross, Alan G. II. Title.
Q223.H37 2010
808'.0665—dc22

2009022745

♾ The paper used in this publication meets the minimum requirements of the American National Standard for Information Sciences—Permanence of Paper for Printed Library Materials, ANSI Z39.48-1992.

CONTENTS

WHAT THIS BOOK DOES

. . . There is nothing more necessary for promoting the improvement of Philosophical Matters, than the communicating to such, as apply their Studies and Endeavours that way.

Henry Oldenburg (1665)

CRAFT. An art, trade, or profession requiring special skill and knowledge . . . sometimes applied to any business, calling, or profession by which a livelihood is earned.

Oxford English Dictionary

Scientific communication proper is a *craft* of which very few men or women become masters by intuition.

Adapted from A. W. Ward (1882)

There are many tens of thousands of scientific periodicals and proceedings. As a result, scientists can find some place that will publish almost anything they write, even when the quality of the writing and of the arguments they make leaves much to be desired. Thus some scientists may legitimately contend: "I can publish whatever I wish. Why should I go to the considerable effort of trying to make it clear as long as the science is sound and the meaning more or less decipherable? It would be a waste of my valuable time. I am a scientist, not a professional writer." There is an unfortunate element of truth to this position. Amidst the bounty of publications, however, attracting the serious interest of anyone is extraordinarily difficult. Thousands of voices cry for attention; only a few are lucky enough to receive it. Achieving that goal, we contend, is more likely if you write prose that creates no serious barriers between your readers and the persuasive argument you have crafted.

Our book is meant as a guide for helping you attain that goal by learning from contemporary scientists who have reached the highest level of achievement in their profession. Indeed, we initially thought to call it *How the Best Scientists Write.* But speaking is also an important activity of scientists, as our chapters on PowerPoint demonstrate. Moreover, constructing a persuasive argument in science is not just a matter of words; tables, graphs, diagrams, and photographs are also vehicles of scientific communication. Indeed, one of the features that distinguish scientific communication from its more literary brethren is this heavy reliance on the visual. Scientific meaning is routinely a product of verbal-visual interaction. Our book thus covers not only the creation of scientific visual displays but also their integration into a coherent narrative and convincing argument.

We divide the book into three parts. Part I examines and exemplifies the distinct sections of the typical scientific article. In the order in which they are presented to their publics, these sections are title, byline, abstract, introduction, methods and materials, results, discussion, conclusion, acknowledgments, and references. But scientists seldom write in this perfectly logical order. When a single author writes an article, the abstract and title may be written last, after the exact content of the body is known. The methods and materials may be tackled first, because they require little creative thinking. On articles with more than one author, it is not uncommon to dole out responsibility. One author might do the abstract and introduction; another, the methods and materials; yet another, the results and discussion as well as the conclusions. One or more "authors" might not be involved in the preparation of the initial draft at all. It is also worth noting that scientific articles are seldom written from scratch. Their authors have already written a proposal or research plan to secure the needed funds, maintained a notebook or computer file to keep track of daily progress, generated many tables and figures of data and observations, and arrived at some conclusions (at least tentative ones) from having analyzed this information and tested their reasoning and conclusions against the skepticism of an invisible college of colleagues. Most of the time, writing an article involves several authors shaping these diverse materials into a convincing, unified argument. Given the existence of these community practices, we decided to organize part I to reflect not the order of typical presentation but one plausible order of composition.

What impels an individual or group to stop experimenting and to start writing? Obviously, the sense that they have solved a scientific problem on the research front. Accordingly, we will start part I with those sections of the article designed to alert potential readers that a problem of interest to the field has been mastered: the introduction, abstract, and title. The introduction and the abstract work together to place the authors' efforts at the leading edge of the appropriate research front, to point to the problem that their research was intended to solve, and to suggest or summarize its solution. Indeed, the abstract and the introduction so overlap in purpose that in short articles (sometimes called "letters") the introduction is the abstract. The title, the final product of this introductory phase, is created from key words in the abstract. It is, as it were, an abstract of the abstract.

We next turn to the article's argumentative core—the material designed to convince skeptical readers that the author's claim to have solved a research problem is credible. Having decided based on the introductory sections that an article warrants further scrutiny, such readers will want to know the results its authors achieved in the lab, office, or field. They will also want to know the authors' take on the significance of these results, and their conclusions. Finally, they will want to learn about the materials and methods by which these results were achieved.

The sections distributing credit—the byline, acknowledgments, and references—can be written at almost any time. We discuss them in the subsequent chapter, which brings to a close our discussion of the individual parts in a typical scientific article.

The last two chapters in part I examine the arrangement of the parts to form a whole. In the first, we present a model of a typical scientific article to illustrate how all these parts work together to communicate a new knowledge claim and argue for its validity. But different content or a different audience for a particular scientific journal can dictate a modest or even substantial variation on the standard model. The concluding chapter in part I guides you through four important variants: a clinical medicine article, theoretical article, literature review, and digital article.

The life of working scientists is not restricted to doing research and publishing in specialist journals. Scientists must also propose new research for funding, publicize their successes to broader audiences, and make presentations at scientific conferences and seminars. Few contemporary researchers rise to the top of their fields without mastering these different forms of scientific communication. These are the topics in part II.

We begin part II with a chapter on research proposals. Of the three types of communication beyond the scientific article, skill in writing proposals is probably the most important to career survival. Almost all scientific papers are preceded by a proposal for funding. All good proposals have the essential elements of a good introduction (establishing an active field of research, defining a limited problem within that field, presenting a possible solution). But the emphasis is different, first, because the work is not yet completed and, second, because the audience is wider. True, the primary audience will be a few scientists with expertise in the field who are looking to fund projects that have a legitimate chance of advancing the research front. But the audience for a given proposal will also likely include a few managers or administrators with scientific training but no such expertise. They are looking to fund projects whose success will advance their careers within an organization whose priorities shift from year to year. Successful research proposals bridge the gap between these two audiences.

The next chapter turns from communicating science to other scientists to communicating science to a lay audience. While writing a good scientific article involves constructing a sound scientific argument, writing about science for a lay audience involves telling a good story. This change in purpose forces scientists to rethink their research, to reframe it as a narrative of wide relevance. To help scientists shift gears, we compare the presentation of science in different periodicals aimed at a general audience. From these comparisons we derive some principles for a successful translation from the expert reader to the general public.

The last two chapters in part II treat PowerPoint, an important current technology for conveying original science to diverse audiences. The first PowerPoint chapter treats the design of individual slides that integrate the verbal and visual into a single thought, free from distractions (what Edward Tufte calls "chart junk"). The following chapter shows how to link such slides into a coherent narrative for a general audience or a convincing argument for a more specialized one. PowerPoint is a powerful but dangerous tool for oral scientific communication. If used properly, it is by far the best means for speakers to communicate their science to diverse audiences. Improperly used, however, it can bore audiences to distraction and distort the science the speaker is trying to convey. We attempt to show how to exploit PowerPoint's strengths and avoid its pitfalls.

We devote part III to a brief treatment of writing style. First, we describe the chief characteristics of current scientific English. These consist of the extensive employment of technical terminology, accompanied by two overriding principles of sentence construction: the systematic use of the passive voice and of complex noun phrases. We then present a strategy for attaining clarity and cogency within these constraints. In part III, we show you how to become a critical reader of your own writing.

We close each chapter but one with practice exercises. As with any skill, practice makes perfect, or at least better. As a rule, we follow each exercise with our answer. Unlike questions in math or science textbooks, however, most of our exercises do not have one and only one correct answer. Your answer may be correct though it differs from ours.

Why bother with our book? If you google "scientific and technical style guides," you will discover that there are already many such books on the market, including the University of Chicago Press's own *Guide to Communicating Science* by Scott Montgomery (2003). This and other similar books offer sound advice on the process involved in moving from idea to draft to submitted manuscript to printed paper. They cover such practical matters as taking notes, constructing an outline, using the Internet, dealing with editors and referees, applying for a research grant, and presenting a scientific paper at a meeting. Some guides also delve into the innumerable stylistic niceties necessary for consistency. Should a space be inserted between a number and the temperature unit °*C*? Should the volume number in a reference be bold or italicized? Does a period go at the end of a reference? What letters should be capitalized in a title or figure caption?

Our book offers something different. To start with, we base our chapters on how good scientists actually write—not how we think they ought to write. We do so by two principal means. First, our advice stems from our research into the development and structure of the scientific paper and on similar research by others. Second, throughout the book we illustrate our points by drawing upon copious examples extracted from actual documents by successful scientists working in many disciplines.

Our earlier research appeared in book form under the title *Communicating Science: The Scientific Article from the 17th Century to the Present* (Gross, Harmon, and Reidy 2002). In that book, we report how scientist-authors communicated their research results in significant journals from the beginning of modern science to the present. Drawing upon this research, we are able to base our principles in the present book on how the best scientists actually communicate. We also rely upon similar research by others. Especially influential for us was the groundbreaking research into scientific communication by John Swales, Charles Bazerman, and M. A. K. Halliday.

We further draw upon advice in how-to-write books and articles we respect. Particularly influential was "The Science of Scientific Writing," an article in *American Scientist* by George Gopen and Judith Swan (1990). Like Gopen and Swan, we owe a considerable debt to the published work on expository writing of the late Joseph M. Williams. A revered professor of English and linguistics for many years at the University of Chicago, Williams wrote several outstanding writing guides, both on his own and with collaborators. The most widely known is *Style,* which in the course of its long and prosperous life has appeared in many editions under the imprint of several publishers, one of which was the University of Chicago Press (1990). Williams's books attracted us because their advice stems from research in applied linguistics and cognitive science; they are based on how expert readers really read and how good writers really write. Unlike the usual run of writing guides, they do not simply rephrase advice dispensed in earlier books. Instead, they transform past research into present practice. And their main concern is not stylistic surface features like the alleged difference between *which* and *that,* the correct use of *hopefully,* and the niceties of the serial comma. Instead, these pioneering guides develop and describe an easy-to-follow method for writing clear and concise sentences, coherent paragraphs, and well-argued articles.

We also learned a great deal from Edward Tufte's *The Visual Display of Quantitative Information* (1983) and *The Cognitive Style of PowerPoint* (2003), William Cleveland's *The Elements of Graphing Data* (1985), and Gunther Kress and Theo van Leeuwen's *Reading Images* (1996). In our view, the works of Tufte, Cleveland, and Kress and van Leeuwen have done for visual communication what Williams's books have done for verbal communication. We hope that, as a consequence, our visual advice gets to the heart of what makes scientific visuals and tables work effectively to help in making the strongest possible arguments.

In writing our book, we mulled over how best to exemplify the verbal and visual advice we offer. We rejected the use of fabricated examples, which can be misleadingly simple. One can easily concoct a sentence that can be improved by switching the verb from passive to the active voice or converting one long convoluted sentence into two shorter and simpler ones. But as we shall later see, not all sentences need be short or in the active voice for good scientific communication. Specially concocted examples can seriously misrepresent actual practice.

Another common strategy we rejected is quoting published examples of "bad" writing and "improving" them: first the readers see a bloated and baggy case, then they see it fit and trim after a strict regimen of dieting and exercise in accord with the authors' version of Weight Watchers for Writers. That strategy poses numerous problems—not least of which is greatly offending the person or persons being criticized, who might legitimately claim that they were quoted out of context. More significant, this practice conveys a serious misapprehension about scientific communication: that it routinely fails to communicate. The remarkable progress of science over the centuries would seem to belie such a conclusion.

We ultimately settled on favorably quoting prose passages or reproducing visual images and PowerPoint slides drawn from "the best scientists," those who publish regularly in the most important journals in their fields and draw enthusiastic audiences at conferences. A serious expository problem here, of course, is that we hijack most of our examples from sources originally intended for a highly specialized audience. Where we think necessary, therefore, we supply contextual information so you may better understand how those selected extracts exemplify the principle being articulated.

For the actual examples of scientific articles, we dipped into the large collection we gathered in researching our second book, *The Scientific Literature: A Guided Tour* (2007), plus the classic *Nature* articles reproduced in the collection *A Century of Nature: Twenty-one Discoveries That Changed Science and the World,* edited for the University of Chicago Press by Laura Garwin and Tim Lincoln (2003). We also searched the Web for stellar articles by distinguished researchers working in disciplines not covered by those two books. For articles aimed at general audiences, we used *American Scientist* and *Scientific American.* For PowerPoint presentations and grant proposals, we relied on the kindness and generosity of scientists who wrote for these magazines.

These are our models, and they or similar ones chosen from your own discipline ought to be yours. Learning from and following the lessons of these models can reduce any anxiety you may feel when faced with the terrors of the blank computer screen—terrors all authors feel. And while this book cannot teach you how to think in a way that will win a Nobel Prize, it can teach you how to use writing and visual display as tools for the forging and shaping of ideas and for conveying these clearly and concisely as you communicate your discoveries to your fellow scientists and fellow citizens.

PART I

The Scientific Article

1 Introducing Your Problem

In a seventeenth-century article, the gentleman scientist Robert Boyle opened with the following brief introduction to his experiments on the respiration of animals:

> Nature having, as *Zoologists* teach us, furnished *Ducks* and other water-Fowl with a peculiar structure of some vessels about the heart, to enable them, when they have occasion to Dive, to forbear for a pretty while respiring under water without prejudice: I thought it worth the tryal, whether such Birds would much better than other Animals endure the absence of the Air in our exhausted Receiver. The accounts of which tryals were, when they were made, registered as follows.

Since Boyle's time, introductions have evolved. Modern readers expect more information than Boyle gives—at least something about recent research on the subject bolstered by apt citations, and maybe a hint at the conclusion from the "tryals." While the science in the best modern scientific articles is never conventional, over time science has adopted a conventional form for the introduction that relies less on the stylistic artistry of individual authors and more on their ability to manipulate three simple components. The subject of this chapter is what those components are and how to bend them to your own expository purposes.

The Structure of the Typical Scientific Introduction

In his classic *Rhetoric,* Aristotle makes the obvious point that all introductions, whether a prologue to a poem, a prelude to a musical composition, or a preface to a speech, pave the way for what is to follow. According to linguist John Swales, modern scientific introductions conventionally accomplish this purpose in three stages:

1. Define a research territory. This stage normally summarizes the state of knowledge in the scientific research front being studied.

2. Establish a limited problem in that territory, one at the leading edge of a research front. In this stage, authors point out a contradiction or inconsistency or gap in that state or propose to build upon a neglected, undeveloped, or misunderstood aspect of it.
3. Suggest or summarize your solution to this problem. This stage typically focuses on the solution to the problem or an approach for solving it. It might also deal with why we should care. In long articles, scientist-authors sometimes end the introduction with a roadmap to the rest of the article. The first two stages of the introduction provide a context for the third. In so doing, they tell readers in what way the conclusions represent a significant contribution to new knowledge: readers are being told why they should read on.

Imagine you are a newly minted child psychologist. You want to conduct a research study of children in a social setting. "How children behave in the presence of others" is too vague a question to begin a research project. Would you be addressing any kind of behavior whatsoever? Who else would be present? What would the participants be doing in the setting? After further deliberation, you decide you want to study children imitating aggressive adults, a topic of no small concern to society at large. Your research territory has narrowed considerably. But that is not enough. You must first know what others have published on this topic. You do not want to reinvent the wheel; if you did, the resulting manuscript would likely be rejected for publication. Even if not, should a priority conflict arise after publication, you would lose. So you google "aggressive behavior," "children," and "adult models" and also search relevant publication databases.

From your search, you discover that others have reported that when children watch an adult behave aggressively with a doll, they imitate the same behavior in the same setting with the model adult present. You decide you want to extend that published research, to take it one step further: do such children still behave aggressively in a different setting with no model adult present? And as side issues you will ask: Are boys more prone to such behavior than girls? And are male adult models more likely to induce aggressive behavior than females? You now have in hand some original research problems in behavioral psychology. At a scientific conference during the winter in Hawaii, you check with some senior colleagues just to make sure your new problems really are new and worth pursuing. Encouragement received, you begin thinking about how to solve those problems.

Those are the research problems addressed by Albert Bandura, Dorothea Ross, and Sheila A. Ross in a classic psychology study published in 1961. We quote their first three paragraphs, inserting italicized headings to remind you of the prototypical three components:

> *Research territory*
>
> A previous study, designed to account for the phenomenon of identification in terms of incidental learning, demonstrated that children readily imitated behavior exhibited by an adult model in the presence of the model (Bandura & Huston, 1961). A series of experiments by Blake (1958) and others (Grosser, Polansky, & Lippitt, 1951; Rosenblith, 1959; Schachter & Hall, 1952) have likewise shown that mere observation responses of a model have a facilitating effect on subjects' reactions in the immediate social setting.
>
> *Problem*
>
> While these studies provide convincing evidence for the influence and control exerted on others by the behavior of a model, a more crucial test of imitative learning involves the generalization of imitative response patterns under new settings in which the model is absent.
>
> *Approach to solving problem*
>
> In the experiment reported in this paper children were exposed to aggressive and nonaggressive adult models and were then tested for the amount of imitative learning in a new situation in the absence of the model. According to prediction, subjects exposed to aggressive models would reproduce aggressive acts resembling those of their models and would differ in this respect both from subjects who served as nonaggressive models and from those who had no prior exposure to any models. This hypothesis assumed that subjects had learned imitative habits as a result of prior reinforcement, and these tendencies would generalize to some extent to adult experimenters (Miller & Dollard, 1941).

This selection does not complete the authors' introduction. It continues for several paragraphs, presenting subsidiary research problems along with hypothesized outcomes. By the end, readers are well prepared for the next section—the details of their method for solving their stated problems.

The next time you read a research article by a leading figure in your field of interest, pay particular attention to the introduction. In some form you will likely find the three stages given by Swales: research territory, problem with that territory, some hint as to the solution to the stated problem. They make sense in all learned communications centered on solving a research problem and defining it in terms of current knowledge. You do not need to be fully cognizant of these three stages to write a good introduction, any more than you need to know the mechanics behind walking to walk. But writing is not as natural as walking. And just as intimate knowledge of the mechanics behind many challenging activities from skiing to poker to computer programming can lead to substantial improvements in performance, so too with writing introductions. This knowledge is also helpful in more efficiently reading and critiquing introductions by others. Finally, it is helpful in modified form for writing proposals to win research funds—a topic we cover in chapter 10.

First Stage: Research Territory

Let's look more closely at the first introductory stage, summarizing the state of knowledge in a research territory. Robert Boyle, writing about respiration in animals in the late seventeenth century, formulates a problem that had probably occurred to most zoologists of his day, indeed to any student of nature. But we are left in the dark as to what others had done, if anything, on his particular problem. In contrast, Bandura and colleagues clearly place the reader in the context of their research front. They do so by summarizing appropriate articles whose bibliographic details will appear in the references section at the end.

In the two following short introductory paragraphs, we also see this principle in action. The first is from an article by physicists Lise Meitner and Otto Frisch (1939). The authors begin by summarizing recent experimental research on what happens when you bombard the heavy element uranium with neutrons:

> On bombarding uranium with neutrons, Fermi and his collaborators[1] found that at least four radioactive substances were produced, to two of which atomic numbers larger than 92 were ascribed. Further investigations[2] demonstrated the existence of at least nine radioactive periods, six of which were assigned to elements beyond uranium, and nuclear isomerism had to be assumed in order to account for their chemical behavior together with their genetic relations. . . . Following up an observation of Curie and Savitch,[3] Hahn and Strassman[4] found that a group of at least three radioactive bodies, formed from uranium under neutron bombardment, were chemically similar to barium and, therefore, presumably isotopic with radium. Further investigation,[5] however, showed that it was impossible to separate these bodies from barium . . .

The superscript numbers of references in the first sentences of this introduction are a tribute to the cumulative achievement that is the essence of an advancing science, in this case, nuclear physics in the 1930s. These sentences lay the intellectual foundation for the authors' theoretical explanation to follow, their theory of nuclear fission, the splitting of uranium atoms into much smaller elements.

How far back in history should these introductory references go? References seldom exceed ten years of age. For example, here is a heavily referenced introductory paragraph by Nobel laureate David Baltimore (1970) on the subject of viruses and infection:

> DNA seems to have a critical role in the multiplication and transforming ability of RNA tumor viruses.[1] Infection and transformation by these viruses can be prevented by inhibitors of DNA synthesis added during the first 8–12 h after exposure of cells to the virus.[1–4] The necessary DNA synthesis seems to involve the production of DNA which is genetically specific for the infecting virus,[5,6] although hybridization studies intended to demonstrate virus-specific DNA have been inconclusive.[1] Also, the formation of virions by the RNA tumor viruses is sensitive to actinomycin D and therefore seems to involve DNA-dependent RNA synthesis.[1–4,7] One model which explains these data postulates the transformation of the infecting RNA to a DNA copy which then serves as a template for the synthesis of viral RNA.[1,2,7] This model requires a unique enzyme, an RNA-dependent DNA polymerase.

Baltimore's six sentences contain sixteen citations to seven sources, none older than seven years. In principle, Baltimore's introduction could easily have extended much further in time: to the discovery of the principle of inheritance, to the first detection of genes, DNA, and RNA, or to the unraveling of the structure of DNA and RNA and their roles in forming proteins. But experts in the biomedical field would already know that history well. Baltimore touches upon only pertinent earlier published theories and experiments necessary for his readers to appreciate the significance of the problem or question addressed in the subsequent paragraph. Indeed, a common failing of scientific introductions is that they give far more information related to the first component than the reader needs to appreciate the forthcoming problem.

Second Stage: Research Problem

The second introductory stage presents the problem the author's research will solve. Scott Montgomery (1996) puts his finger on the nature of these well-formed problems within subdisciplines when he speaks of the job of researchers as "an unending attempt to create the conditions for new work, to find gaps or instabilities in existing [intellectual] structures." The first stage only paves the way for the second. Without the problem, there would be no research to report.

Research problems arise from a variety of sources. In the introduction mentioned earlier, for example, Baltimore (1970) argues that something is missing in current molecular biology—its processes require an as-yet-undetected enzyme:

> No enzyme which synthesizes DNA from an RNA template has been found in any type of cell. *Unless such an enzyme exists in uninfected cells,* the RNA tumor viruses must either induce its synthesis soon after infection or carry the enzyme into the cell as part of the virion [our emphasis].

In another example, Meitner and Frisch (1939) formulate an as-yet-unanswered question: Why does bombarding uranium with neutrons produce a much smaller element, in contradiction to current theory?

> . . . Hahn and Strassmann were forced to conclude that isotopes of barium (Z [atomic number] = 56) are formed as a consequence of bombardment of uranium (Z = 92) with neutrons.
>
> *At first sight, this result seems very hard to understand.* The formation of elements much below uranium has been considered before, but was always rejected for physical reasons, so long as the chemical evidence was not entirely clear cut [our emphasis].

In our third example, Joseph Farman and colleagues (1985) uncover an alarming inconsistency in current data on the ozone layer in the Antarctica stratosphere:

> Thus, two spectrophotometers have shown October values of O_3 to be much lower than March values, a feature entirely lacking in the 1957–73 data set. To interpret this difference as a seasonal instrumental effect *would be inconsistent* with the results of routine checks using standard lamps [our emphasis].

Stage 2 is sometimes not so much a problem as a need for a better tool to solve problems. In their introduction, Erwin Neher and Bert Sakmann (1976) establish the need for a superior method of measuring current flow across biological membranes so that scientists can study the electrical activity in nerve synapses:

> Clearly, it would be of great interest to refine techniques of conductance measurement in order to resolve discrete changes in conductance which are expected to occur when single channels open or close. *This has not been possible so far because of excessive extraneous background noise* [our emphasis].

From these examples we may infer, correctly, that the typical scientific article attempts to solve only well-formed problems within subdisciplines or to suggest new methods for solving these problems. Such problems do not directly concern pressing societal imperatives such as whether the earth's atmosphere is being depleted of ozone; they do not tell us how to eradicate disease or solve the energy crisis. Rather, they concern something far more limited: Why is there an apparent anomaly in the ozone measurements over Antarctica? Does this new drug significantly increase the survival chances of breast cancer patients compared with the old treatment or no treatment at all? Does the electrochemical reaction running at room temperature in this new tabletop device emit neutrons at a rate consistent with nuclear fusion? What is the viscosity of liquid helium below the λ-point?

Scientists fashion such problems out of the inconsistent claims or gaps in knowledge within their disciplines. The resolution of these problems is usually anticipated by a clue—a word or turn of phrase indicating a gap or inconsistency, italicized in the selections above. A common failing is to bury the problem in such a way that the reader finishes the introduction without understanding what exactly the problem is.

Third Stage: Solution Forthcoming

You might wonder how to proceed with stage 3 if, at so early a stage in the writing process, you have not yet reached a firm conclusion about your research. In that case, you might want to formulate a provisional conclusion and return to that formulation later, after having completed your first draft of the core sections (results, discussion, and conclusion). Long before writing commences, however, most authors of scientific manuscripts possess at least a rough idea of their main conclusion. This is especially true when, as is often the case, the research on which the article is based has been a consequence of a grant application.

Here are two examples that answer this third question. Baltimore (1970) uses a single sentence to summarize his solution to the problem concerning the mechanism of tumor viruses:

> This study demonstrates that an RNA-dependent DNA polymerase is present in the virions of two RNA tumor viruses: Rauscher mouse leukaemia (R-MLV) and Rous sarcoma virus.

The introduction to the Neher and Sakmann article (1976) announces their application of a new biochemical method to frog muscles:

> We report on a more sensitive method of conductance measurement, which, in appropriate conditions, reveals discrete changes in conductance that show many of the features that have been postulated for single ionic channels.

A classic astronomical article provides a more elaborate example of stage 3. In 1967, astronomers at the Mullard Radio Astronomy Observatory near Cambridge, England, detected a highly unusual pulsing signal beaming from outer space. Their problem was simple: how to explain it. Could it have been a signal from some advanced extraterrestrial civilization? That thought was seriously entertained at first. The eventual solution proposed by Antony Hewish and his collaborators (1968) was not quite that spectacular but still a major achievement:

> The remarkable nature of these signals at first suggested an origin in terms of man-made transmissions which might arise from deep space probes, planetary radar or the reflexion of terrestrial signals from the Moon. None of these interpretations can, however, be accepted because the absence of any parallax shows that the source lies far outside the solar system. A preliminary search for further pulsating sources has already revealed the presence of three others having remarkably similar properties[,] which suggests that this type of source may be relatively common at a low flux density. A tentative explanation of these unusual sources in terms of the stable oscillations of white dwarf or neutron stars is proposed.

Obviously, this paragraph is more than a brief solution statement. In part, it is a preliminary argument in favor of the authors' solution. The Hewish group anticipates solutions their readers might initially propose. They first tell us they ruled out several man-made sources within our solar system. They then mention that continued searching of the heavens has indicated that pulsing radio sources appear to be fairly common. If the sources are many, it is implied, they cannot be sent by extraterrestrial beings. Finally, we are given a proposed solution—dense collapsed stars that spin and emit pulsating signals, entities that were eventually named *pulsars*.

Alternative Introductory Structures

The content and order of the three stages identified by Swales are not meant as rules to live by religiously but as basic building blocks out of which writers can craft infinite variations. A popular alternative strategy is to pique reader interest by beginning with a solution statement. In their DNA article of 1953, for instance, James Watson and Francis Crick open with a straightforward and understated announcement of their discovery: "We wish to suggest a structure for the salt of deoxyribose nucleic acid (D.N.A.)." And in his $E = mc^2$ article of 1905, Albert Einstein uses similar plain language to announce a discovery he made concerning an implication of the relativity principles set forth in his earlier 1905 article: "The results of the previous investigation ['On the electrodynamics of moving bodies'] led to a very interesting conclusion, which is here deduced."

Scientific introductions can also perform functions other than introducing a research problem and its solution. One is providing a roadmap for the remainder of the article. Our example is the penultimate paragraph from the introduction to a sixty-page *Nature* article by the International Human Genome Sequencing Consortium (2001). This paragraph presents the article's organization in a nutshell:

> In this paper, we start by presenting background information on the project and describing the generation, assembly and evaluation of the draft genome sequence. We then focus on an initial analysis of the sequence itself: the broad chromosomal landscape; the repeat elements and the rich palaeontological record of evolutionary and biological processes that they provide; the human genes and proteins and their differences and similarities with those of other organisms; and the history of genomic segments. . . . Finally, we discuss applications of the sequence to biology and medicine and describe next steps in the project.

We would not recommend this sort of paragraph for the typical scientific article. But for long articles that depart from the conventional overall structure, such a paragraph orients the readers to the subject matter that follows—making it easier for them to concentrate on the material of personal interest. When you are being taken on a long trip, it is comforting to know where you are going and by what route you are going to get there.

So What? Why Should We Care?

In a parody of the typical scientific paper ("The Super G-String"), physicist Warren Siegel (1986) begins with the following anti-introduction:

> Actually, this paper doesn't need an introduction, since anyone who's the least bit competent in the topic of the paper he's reading doesn't need to be introduced to it, and otherwise why's he reading it in the first place? Therefore, this section is for the referee.

It is, unless you realize that introductions are meant to entice. Readers are less likely to ignore your article if they are told up front why they should care that you solved this particular problem. To achieve that end, we recommend the "so-what" technique—that is, we recommend that you think about your research problem and its solution, put yourself in the shoes of your potential reader, and ask, "So what? Why should readers care?"

As an example, let's turn again to Watson and Crick's famous article on the structure of the DNA molecule. What makes the article famous, of course, is not the discovery of the structure alone but the discovery that the structure suggested a copying mechanism for genetic transfer. To a reader who might have said, "You discovered the structure of some molecule—so what?" Watson and Crick (1953) responded in the second sentence of their two-sentence introduction: "We wish to suggest a structure for the salt of deoxyribose nucleic acid (D.N.A.). This structure has novel features which are of considerable biological interest." Watson and Crick do not tell us what they mean by "considerable biological interest" until their short article's closing paragraphs.

Here is another introduction, one that handles the so-what question in a little more expansive fashion. It is from an article concerning the effects of genetic mutations on the development of the fruit fly (*Drosophila*), written by Christiane Nüsslein-Volhard and Eric Wieschaus (1980). The first paragraph reads as follows:

> *Research territory*
> The construction of complex form from similar repeating units is a basic feature of spatial organization in all higher animals.
>
> *Problem*
> Very little is known for any organism about the genes involved in the process.
>
> *Solution*
> In *Drosophila,* the metameric nature of the pattern is most obvious in the thoracic and abdominal segments of the larval epidermis[,] and we are attempting to identify all loci required for the establishment of the pattern. [*So what?*] The identification of these genes and the description of their phenotypes should lead to a better understanding of the general mechanisms for the formation of metameric patterns.

We need not stop our questioning with a single "so what?" as these authors do. Let's continue on our own.

So what? This work could lead to an understanding of how complexity arises out of simplicity in primitive biological organisms like fruit flies.

So what? It could also, in the very distant future, lead to a detailed understanding of the genes controlling embryo development in other animals, maybe even humans.

So what? That understanding might be used one day in treating genetically inherited diseases.

As we traverse the so-what chain, please note, we move from the specific problem or problems being researched, to disciplinary goals, to societal benefits or potential societal problems down the road. To a great extent, where to stop in this chain depends on your intended audience. If your audience is only or mostly those working on the same or similar research problems, you need not elaborate disciplinary goals. In fact, our impression is that most scientific introductions stop well short of broader claims. If, however, the projected audience includes others trained in your discipline but not working on the same kinds of problems, you might want to link your problem to broader disciplinary goals, just as Nüsslein-Volhard and Wieschaus do. And if your research has potential news value for the general public and might be read by science journalists, then any societal benefits or problems may be *cautiously* mentioned. We use the emphasis here because we recommend that if these possible benefits or dangers are not imminent, you are normally better off not wading into those waters. There is a real danger to overclaiming or misleading readers.

In any case, reference to broader goals may be more appropriate in the concluding section of your article. See chapter 5.

Making a Favorable First Impression

In any literary document, the first sentence takes on heightened importance because, well, it comes first, and some busy readers might not go much beyond that point because that sentence fails to engage them. "I hate traveling and explorers." Thus begins the 1955 book *Tristes tropiques* (untranslatable, but literally meaning "sad tropics") by the French anthropologist Claude Lévi-Strauss. Lévi-Strauss's memoir describes his experiences traipsing through dense Brazilian jungle in the 1930s while exploring kinship relationships and other social structures within the Amerindian tribes he encountered. Our first reaction to Lévi-Strauss's first sentence is, Why would someone whose profession is largely taken up with traveling and exploring make such an opening claim? The book that follows it tells us why—and much more besides. It is one of the great travelogues from the twentieth century, as well as a seminal contribution to the anthropological literature.

Clearly, that catchy first sentence signals that we are in the hands of a very skilled prose stylist. Indeed, we generally expect such masterful first sentences in poems, novels, and memoirs. No such expectation exists for modern scientific articles. Still, some scientists are not without a certain flair in the creation of enticement. For the fun of it, see if you can correctly match the first sentences on page 14 with the names of the famous scientists who composed them.

. . . in the beginning of the Year 1666 (at which time I applyed my self to the grinding of Optick glasses of other figures than Spherical,) I procured me a Triangular glass-Prisme, to try therewith the celebrated Phenomena of Colours.	Albert Einstein (1905b)
In this volume I have attempted to expound the methods and results of dream-interpretation; and in so doing I do not think I have overstepped the boundary of neuro-pathological science.	Richard Feynman (1965)
This paper represents a first crude effort to explain the large-scale features of the earth's surface, that is, the continental masses and the ocean basins.	Linus Pauling et al. (1951)
Determinations of the motion of the sun with respect to the extra-galactic nebulae involved a K term of several hundred kilometers which appears to be variable.	Marie Curie (1898)
It is known that Maxwell's electrodynamics—as usually understood at the present time—when applied to moving bodies, leads to asymmetries which do not appear to be inherent in the phenomena.	Sigmund Freud (1911)
During the past fifteen years we have been attacking the problem of the structure of proteins in several ways.	Isaac Newton (1672)
We have a habit in writing articles published in scientific journals to make the work as finished as possible, to cover all the tracks, to not worry about the blind alleys or to describe how you had the wrong idea first, and so on.	Alfred Wegener (1912)
I have studied the conductance of air under the influence of the uranium rays discovered by M[onsieur]. Becquerel, and I examined whether substances other than compounds of uranium were able to make the air a conductor of electricity.	Edwin Hubble (1929)

Answers, in order from top to bottom: Newton, Freud, Wegener, Hubble, Einstein, Pauling, Feynman, Curie

Good first sentences in scientific articles establish a research territory, disciplinary goals, specific problems, or forthcoming solutions in language that is as straightforward as possible. They do not leave readers feeling as if they have walked into the middle of a conversation in a foreign language. They encourage readers to learn more. For the first sentence quoted earlier from the Watson and Crick DNA article, if we were biochemists in the 1950s, we would eagerly want to find out what its biochemical structure is. For the Einstein $E = mc^2$ article, if we were physicists at the turn of the century who had read and admired his revolutionary first article, we would be dying to learn about the "interesting conclusion" derived from his continued study.

Conclusion

This chapter began with the three-step introduction derived by John Swales from a linguistic analysis of research articles. His analysis is descriptive: it tells us what contemporary scientists actually do when they write an introduction. But his analysis also has normative implications: it tells us what scientists *should normally do* when they write introductions. The point we would like to make in closing is that you can convert Swales's template into a set of questions that can help decide whether your research is ready to be shared with colleagues in the form of a research article. This readiness depends, we think, on clear answers to the following: What is your research territory? How have you limited that territory so that a specific problem emerges, one that is at the leading edge of the research front? And—most important—what do you think might be your solution to this problem? Will the solution relate to something that others in your field will want to know about? In short: why do you think, having read your introduction, readers should read on? Having answered those important questions at least tentatively, you will be better prepared to write not only a good introduction but the remaining article sections as well.

EXERCISES

To repeat: good introductions put readers in the picture, tell them why an addition or alteration to the picture is necessary, and anticipate what the consequences of that addition or alteration will be. The grateful reader exits with a clear understanding of the problem being tackled while having been briefed with sufficient background information to appreciate its historical context and importance. Here are a couple of exercises for honing your introduction-writing skills.

Exercise 1

We extracted the following introductory paragraph from a *Nature* article (Lauterbur 1973) announcing the development of a new imaging method—what later came to be called the "nuclear magnetic resonance" (NMR) method and "magnetic resonance imaging" (MRI). This powerful imaging method is now readily available at U.S. medical centers big and small. We have numbered each sentence. Categorize each as a component of one of the three stages of introductions.

1. An image of an object may be defined as a graphical representation of the spatial distribution of one or more of its properties.
2. Image formation usually requires that the object interact with a matter or radiation field characterized by a wavelength comparable to or smaller than the smallest features to be distinguished, so that the region of interaction may be restricted and a resolved image generated.
3. This limitation on the wavelength of the field may be removed, and a new class of image generated, by taking advantage of induced local interactions.
4. In the presence of a second field that restricts the interaction of the object with the first field to a limited region, the resolution becomes independent of wavelength, and is instead a function of the ratio of the normal width of the interaction to the shift produced by a gradient in the second field.
5. Because the interaction may be regarded as a coupling of the two fields by the object, I propose that image formation by this technique be known as zeugmatography, from the Greek ζεγμά, "that which is used for joining."

Answers

Research territory: sentence 1
Research problem: sentence 2
Solution: sentences 3–5

Note that the final sentence is only tangentially related to the solution; its main point is to name the new technique described in the main text. (The name did not stick.) The point here is that the introduction can easily handle topics other than the prototypical three we defined at the beginning of this chapter.

Exercise 2

For the next exercise, we have extracted two introductory paragraphs from a psychology paper (Bandura, Ross, and Ross 1963) on how watching violent actions in films affected the behavior of young children:

> Most of the research on the possible effects of film-mediated stimulation upon subsequent aggressive behavior has focused primarily on the drive reducing function of fantasy. While the experimental evidence for the catharsis or drive reduction theory is equivocal (Albert, 1957; Berkowitz, 1962; Emery, 1959; Feshbach, 1955, 1958; Kenny, 1952; Lövaas, 1961; Siegel, 1956), the modeling influence of pictorial stimuli has received little research attention.
>
> In an earlier experiment (Bandura & Huston, 1961), it was shown that children readily imitated aggressive behavior exhibited by a model in the presence of the model. A succeeding investigation (Bandura, Ross, & Ross, 1961) demonstrated that children exposed to aggressive models generalized aggressive responses to a new setting in which the model was absent. The present study sought to determine the extent to which film-mediated aggressive models may serve as an important source of imitative behavior.

In these paragraphs, the authors portray their research as a logical extension of past published research in child psychology and aggressive behavior, including their own work. But how would you feel if the following paragraph were interposed between the two above?

> A recent incident (*San Francisco Chronicle,* 1961) in which a boy was seriously knifed during re-enactment of a switchblade knife fight the boys had seen the previous evening on a televised rerun of the James Dean movie, *Rebel without a Cause,* is a dramatic illustration of the possible imitative influence of film stimulation. Indeed, anecdotal data suggest that portrayal of aggression through pictorial media may be more influential in shaping the form aggression will take when a person is instigated on later occasions, than in altering the level of instigation to aggression.

Answer

As you may have guessed, that last paragraph does indeed appear in the actual journal article. We see nothing wrong with it. In defining a research territory you need not stick rigorously to the published scientific literature. You can weave in anecdotes, newspaper accounts, e-mail, or whatever you want if it helps to introduce your research problem. In this particular case, the authors exploited such evidence to formulate a hypothesis worth putting to the test.

CHECKLIST

Now examine one of your own past or present introductions:

- Does it have all three stages in the typical order?
- If not, does there appear to be a good reason why a stage is missing or out of the conventional order?

- Does the intellectual context adequately set the stage for the problem statement?
- Will the reader be persuaded that you have consulted the key references relevant to your research problem?
- Does the research problem clearly stand out from your review of the existing literature?
- Is the problem statement specific and focused?
- Is the solution, or your approach to solving it, mentioned at least briefly?
- Have you asked yourself why readers should consider solving the problem to be a worthwhile endeavor?
- If the article is very long, would the reader benefit from a short closing passage outlining the entire article's contents?

2 Distilling Your Research

In a 1980 issue of the *Journal of Immunology*, B. H. Waksman exaggerated only slightly when he editorialized that "the real articles are the abstract; the rest . . . is a technical report, available if needed, but rarely read." Informative abstracts—the essence of a much longer work's contents—have existed at least since the dawn of the scientific periodical in the seventeenth century. But only in the midtwentieth century did they become standard fixtures atop individual scientific articles. Einstein's four famous papers of 1905 had none. What is the purpose of informative abstracts? Why have they flourished over the last century?

Creating Informative Abstracts

Here is an example abstract by Peter T. Boag and Peter R. Grant (1981) on Darwin's finches in the Galápagos Islands:

> Survival of Darwin's finches through a drought on Daphne Major Island was nonrandom. Large birds, especially males with large beaks, survived best because they were able to crack the large and hard seeds that predominated in the drought. Selection intensities, calculated by O'Donald's method, are the highest yet recorded for a vertebrate population.

Informative abstracts such as the above have flourished because they rapidly answer the overarching question "If I read this article, what will I learn?" Moreover, in the best examples they accomplish this task without the reader's having to refer to the article itself. Boag and Grant's abstract does so by answering three basic questions in three sentences:

1. *What was done?* Survival rates of Darwin's finches were measured during a drought on a Galápagos island.
2. *How was it done?* O'Donald's method was used to measure selection intensity during the drought.
3. *What was discovered?* The selection intensity was revealed to be a record high.

Armed with that information and the article's title, most evolutionary biologists would want to learn more. This basic structure is standard practice for abstracts in all disciplines that involve applying some repeatable quantitative method to solving a research problem. We will examine two more examples: one from behavioral psychology, one from molecular biology.

An article by Bandura, Ross, and Ross (1963) reports an experiment in which children were divided into experimental groups and subjected to films that portrayed aggressive behavior. It answers those same questions in a little more detail, especially the second question:

> *What was done*
> To test the hypothesis that exposure of children to film-mediated aggressive models would increase the probability of Ss' [the subjects'] aggression to subsequent frustration . . .
>
> *How it was done*
> . . . 1 group of experimental Ss observed real-life aggressive models, a 2nd observed these same models portraying aggression on film, while a 3rd group viewed a film depicting an aggressive cartoon character. Following the exposure treatment, Ss were mildly frustrated and tested for the amount of imitative and nonimitative aggression in a different setting . . .
>
> *What was discovered*
> The overall results provide evidence for both the facilitating and the modeling influence of film-mediated aggressive stimulation. In addition, the findings reveal that the effects of such exposure are to some extent a function of the sex of the model, sex of the child, and the reality cues of the model.

Our next example is the abstract from a famous *Nature* article titled "Nucleotide Sequence of Bacteriophage ΦX_{174} DNA" (Sanger et al. 1977), where the emphasis is on answering the third question:

> *What was done*
> *A DNA sequence for the genome of bacteriophage ΦX_{174} of approximately 5,375 nucleotides* has been determined . . .
>
> *How it was done*
> using the rapid and simple "plus and minus" method.
>
> *What was discovered*
> The sequence identifies many of the features responsible for the production of the proteins of the nine known genes of the organism, including initiation and termination sites for the proteins and RNAs. Two pairs of genes are coded by the same region of DNA using different reading frames [our emphasis].

This abstract opens with an announcement of a solution to an unstated problem: the need for the first ever complete DNA sequencing of an organism's genome—an important problem at the time in molecular biology. The italicized phrase restates the article's title with a little more technical information—namely, the sequencing involved a genome consisting of 5,375 nucleotides; in other words, it belonged to a fairly primitive organism given that the human genome has three billion. At this point we know the authors' central claim: the DNA sequencing of bacteriophage ΦX_{174}'s entire genome. The authors could have ended the sentence there. Instead, they appended the participle phrase beginning with "using." This establishes a secondary but still important claim: that their sequencing was done by a rapid and simple method. It also specifies what that method was. The next sentences elaborate upon what they found from having applied the method. In sum, this genome-sequencing abstract answers the questions "what was done" and "how it was done" in the first sentence; the remaining sentences inform us "what was discovered."

Because of their brevity, abstracts must leave out much information that might also be of interest to readers. Here are a few questions a reader might have after reading the Sanger et al. abstract: Why choose this particular organism to sequence? How far have other researchers gotten in this area? What is the plus and minus method? How do the authors know it yielded accurate results? What is the overlap of pairs of genes all about? Why should we care that they sequenced this particular genome? To learn the answers to those kinds of questions, readers must dip into the main text. Good titles and abstracts tantalize the reader; at the same time, they stand by themselves as summaries of an article's contents.

Creating Informative Abstracts That Persuade

Good abstracts need not only inform about the article's contents; they can also motivate more readers to continue reading by addressing a fourth question, "Why is the discovery important?" Since abstracts are functionally independent of the articles whose content they summarize, and since they are also far more likely to be read than the articles they head, it makes perfect sense for them to contain, in addition to their obligatory summaries, some indication of the importance of the problem solved or the discovery made. It seems worthwhile for scientists to consider adding a sentence or more designed to turn an abstract into a persuasive document. By "persuasive" here we mean intended to convince readers that the discovery has potential real-world application or advances some research front in a significant way.

Let's now revisit the three earlier abstracts. Let's revise them so as to transform them from informative to informative and persuasive, from giving just the facts to signaling the wider importance of those facts. We accomplish this task by paraphrasing sentences from the bodies of the articles in question. Here is a revision of the abstract of Boag and Grant, with our added material set in italic type:

> Survival of Darwin's finches through a drought on Daphne Major Island was nonrandom. Large birds, especially males with large beaks, survived best because they were able to crack the large and hard seeds that predominated in the drought. Selection intensities, calculated by O'Donald's method, are the highest yet recorded for a vertebrate population. *Our results are consistent with the growing opinion among evolutionary ecologists that the trajectory of even well-buffered vertebrate species is largely determined by occasional "bottlenecks" of intense selection during a small portion of their history.*

Here is a revised version of Bandura, Ross, and Ross given the same treatment, but with the importance made apparent through the addition of a context-setting first sentence:

> *Anecdotal data suggest that children's exposure to the portrayal of adult aggression through pictorial media leads to imitation in real life.* To test the hypothesis that exposure of children to film-mediated aggressive models would increase the probability of Ss' [the subjects'] aggression to subsequent frustration, 1 group of experimental Ss observed real-life aggressive models, a 2nd observed these same models portraying aggression on film, while a 3rd group viewed a film depicting an aggressive cartoon character. Following the exposure treatment, Ss were mildly frustrated and tested for the amount of imitative and nonimitative aggression in a different setting. The overall results provide evidence for . . .

Here is a revised version of Sanger et al.:

> *Various attempts at establishing the complete sequence of the genome of bacteriophage ΦX_{174} have met with only partial success. This bacteriophage consists of approximately 5,375 DNA nucleotides coding for nine known genes. In this report, we describe considerable progress in determining the complete DNA sequence for this organism by the rapid and simple "plus and minus" method.* The sequence identifies many of the features responsible for the production of the proteins of the nine known genes of the organism, including initiation and termination sites for the proteins and RNAs. Two pairs of genes are coded by the same region of DNA using different reading frames.

Our revisions are not meant in any way as criticisms of the originals. The first and third examples appeared in *Science* and *Nature,* respectively. Both those distinguished journals severely limit the length of most research articles (under two thousand words). When the main articles are so short, the accompanying abstract must also be short. If the original short abstract entices, the reader can easily skim the article for the sort of information we have added. The point we hope our revisions demonstrate is that when space is not a premium, you should give serious consideration to writing an abstract that both informs and persuades.

Creating Informative Abstracts from Different Content

Different content will dictate a different approach to writing the abstract. The research behind most scientific articles involves generating and analyzing data derived by applying some method or operating some measuring instrument to solve a research problem. But not all. Some authors report on a new or modified theory. Some review the existing literature within some active research front. Some observe and describe nature. Some report a new invention. So how do you write an abstract when "how it was done" is not relevant or particularly important? We suggest you consider the following questions: What was done? What was discovered? Why is it important?

The first example abstract we have extracted from an article reporting a new map design for the universe (Gott et al. 2003):

> We have produced a new conformal map of the universe illustrating recent discoveries, ranging from Kuiper belt objects in the Solar system, to the galaxies and quasars from the Sloan Digital Sky Survey. This map projection, based on the logarithm map of the complex plane, preserves shapes locally, and yet is able to display the entire range of astronomical scales from the Earth's neighborhood to the cosmic microwave background. The conformal nature of the projection, preserving shapes locally, may be of particular use for analyzing large scale structure. Prominent in the map is a Sloan Great Wall of galaxies 1.37 billion light years long, 80% longer than the Great Wall discovered by Geller and Huchra and therefore the largest observed structure in the universe.

The first sentence announces what the authors did: invented a "new conformal map of the universe." The authors do not mention how they invented this map, since the main audience is astronomers and astrophysics, not mapmakers. The remaining sentences elaborate on the first sentence by telling us what this map does differently from previous versions, how it works, and what it is good for. Such abstracts emphasize what new thing the authors achieved.

As a second example, we have chosen a newsworthy *Nature* article by Edward Daeschler, Neil Shubin, and Farish Jenkins (2006). It reports the discovery of an intermediate fossil representing a missing evolutionary link between fish and four-limbed creatures capable of walking:

> The relationship of limbed vertebrates (tetrapods) to lobe-finned fish (sarcopterygians) is well established, but the origin of major tetrapod features has remained obscure for lack of fossils that document the sequence of evolutionary changes. Here we report the discovery of a well-preserved species of fossil sarcopterygian fish from the Late Devonian of Arctic Canada that represents an intermediate between fish with fins and tetrapods with limbs, and provides unique insights into how and in what order important tetrapod characters arose. Although the body scales, fin rays, lower jaw and palate are comparable to those in more primitive sarcopterygians, the new species also has a shortened skull roof, a modified ear region, a mobile neck, a functional wrist joint, and other features that presage tetrapod conditions. The morphological features and geological setting of this new animal are suggestive of life in shallow-water, marginal and subaerial habitats.

Daeschler and company make no mention of how they discovered this amazing fossil or what they did to it in the laboratory to arrive at their solution. But there really is no need for including that information since readers do not need it to appreciate the importance of the problem or the value of the solution. The authors also make no mention in the abstract (or the main text, for that matter) of their discovery's relevance to the current heated debate between evolutionary theory and creationism. This debate is marginal to their important discovery and would only detract from their scientific message. The authors wisely leave that subject for more public forums. While no direct mention of methods appears, this abstract does tell us quite a bit:

- *What was done?* The authors examined the morphological features of a well-preserved fossil of an ancient (sarcopterygian) fish that they recently discovered in Arctic Canada.
- *What was discovered?* The fossil has anatomical features characteristic of both fish and tetrapods.
- *Why is it important?* Their analytical results provide "unique insights into how and in what order important tetrapod characters arose." Who would not want to read more after that enticement?

Avoiding Uninformative Abstracts

We find the four-sentence abstract of Daeschler, Shubin, and Jenkins interesting for what it does not communicate. The authors could have easily taken a different approach to their abstract. For instance, they could have dropped the context-setting first sentence and gone with all conclusion (sentences 2–4):

> Here we report the discovery of a well-preserved species of fossil sarcopterygian fish from the Late Devonian of Arctic Canada that represents an intermediate between fish with fins and tetrapods with limbs, and provides unique insights into how and in what order important tetrapod characters arose. Although the body scales, fin rays, lower jaw and palate are comparable to those in more primitive sarcopterygians . . .

Or they could have dropped the third and fourth sentences and greatly elaborated the introductory information in sentence 1. To illustrate the effect of such changes, we assembled this alternative abstract by copying and pasting sentences from the article's introduction, ending with sentence 2 above from the original abstract, indicated by italics:

> The evolution of tetrapods from sarcopterygian fish is one of the major transformations in the history of life and involved numerous structural and functional innovations. During the origin of tetrapods in the Late Devonian (385–359 million years ago), the proportions of the skull were remodelled, the series of bones connecting the head and shoulder was lost, and the region that was to become the middle ear was modified. At the same time, robust limbs with digits evolved, the shoulder girdle and pelvis were altered, the ribs expanded, and bony connections between vertebrae developed. Few of these features, however, are seen in the closest relatives of tetrapods—the elpistostegalian fishes—which are incompletely known. In view of the morphological gap between elpistostegalian fish and tetrapods, the phylogenetic framework for the immediate sister group of tetrapods has been incomplete, and our understanding of major anatomical transformations at the fish-tetrapod transition has remained limited. *Here we report the discovery of a well-preserved species of fossil sarcopterygian fish from the Late Devonian of Arctic Canada that represents an intermediate between fish with fins and tetrapods with limbs.*

Our experience is that this sort of abstract—top heavy with introductory background information—tends to get written before authors know exactly what they want to say about their own work. That situation often occurs in the case of papers presented at scientific meetings and later published in a proceedings volume, where the meeting organizers requested an abstract long before the full paper had been prepared or the research itself had been finished. But we would discourage this kind of abstract for a journal article. Scientists read abstracts to learn what is new up front, not what is already known.

Integrating the Abstract into the Front Matter

There is much overlap in the information provided in the front matter: title, abstract, and introduction. All three might make some mention of what was done, how it was done, and what was discovered. Yet the purposes of the three should not be confused. Good titles typically distill what was done or discovered into a single thought and hence a single phrase or full sentence. Abstracts expand on that thought in the manner we have just described. In contrast, introductions center on establishing the nature of the research problem to be solved. Take the introduction to the earlier quoted article on Darwin's finches by Boag and Grant, "Intense Natural Selection in a Population of Darwin's Finches (Geospizinae) in the Galápagos":

> There are few well-documented examples of natural selection causing avian populations to track a changing environment phenotypically. This is partly because birds meet environmental challenges with remarkable behavioral and physiological flexibility,[1] partly because birds have low reproductive rates and long generation times, and partly because it has been difficult for ecologists to quantify corresponding phenotypic and environmental changes in most field studies. In this report we demonstrate directional natural selection in a population of Darwin's finches and identify its main cause.

Despite some overlap, a comparison between this article's title and abstract on the one hand and its introduction on the other reveals an important division of communicative labor: while the job of the title and abstract is to inform with a preview of the article's contents, the job of this introduction is to provide intellectual background for the promise of its third sentence, a promise designed to motivate readers to find out more about what is meant by "directional natural selection in a population of Darwin's finches."

Should the First Be Last?

Some writing guides recommend writing the abstract last even though it appears at the front of the printed document. This advice makes perfect sense since the abstract summarizes the contents of the article as a whole, from introduction to conclusion. But in practice, the writing of complex documents seldom proceeds in a rigorously logical order. In *Shaping Written Knowledge*, for example, Charles Bazerman (1988) analyzed the many revisions that went into the composition of a classic 1923 paper related to quantum theory by Arthur Compton. Bazerman found that Compton had written a draft abstract "about two-thirds of the way through the draft of the main text." He surmises that Compton did this "in order to articulate his sense of the whole and to keep the later parts logically and structurally consistent." We also suspect that Compton wanted to preview an abbreviated version of his projected entire article as soon as the inspiration struck. Skilled writer that he was, Compton revised this draft abstract once he had penned his conclusions.

What is the lesson to be learned here? It is that you can draft your abstract as soon as you have a satisfactory grasp of your argument—whether that is near the beginning, in the middle, or toward the end of the composing process. This draft can act as a sort of expanded table of contents, helping you to strengthen your overall argument and guiding you toward the finish line. Of course, you should also follow Compton's example and return to the draft abstract at the very end to ensure that it is consistent with the entire manuscript.

Conclusion

The brevity of a typical abstract belies its importance. Along with the title, a good abstract enables readers to make a rapid, informed decision as to whether the remaining article is worth their time and effort: it gives readers the gist. Poor abstracts leave readers in the dark regarding what the authors did, how they did it, and what was discovered; frustrated readers must turn to the main text to find out that information, if they are so inclined despite an unfavorable first impression.

EXERCISES

Exercise 1

Here are our versions of abstracts for two of the most famous papers published in the twentieth century (neither original has an abstract).

> In an earlier theoretical investigation of the electrodynamics of moving bodies, I developed a principle of relativity in which the laws of any two physical systems moving uniformly with respect to each other depend upon the frame of reference chosen. Continued manipulation of equations based on this principle led to the interesting conclusion that energy in any form lost from a body will decrease its mass by the amount of that energy divided by the speed of light squared. (abstracted from Einstein 1905a)

> Three-chain structures for the genetic material deoxyribonucleic acid have recently been proposed by others but are unsatisfactory for various reasons. We put forward a different structure, consisting of two helical chains

each wound around a common axis. These chains are held together by the purine and pyrimidine bases. This new structure is consistent with the available experimental data and stereochemical arguments. The double helical arrangement makes for a simple copying mechanism with important biological implications. (abstracted from Watson and Crick 1953)

See if you can identify the components of the abstract that appear in our two examples.

Answer

By means of mathematical manipulations (how it was done), Einstein has derived from his earlier theory (what was done) an important conclusion about the interchangeability between matter and energy (what was discovered).

Watson and Crick built a model of the DNA molecule (what was done) that is in the shape of a double helix, is consistent with the evidence, and violates no chemical laws (what was discovered). This structure has important biological implications (why it is important).

Exercise 2

Below we reproduce several paragraphs extracted from the main body of a scientific article that proved to be instrumental in winning the authors a Nobel Prize. We have defined a few of the key technical terms to help you along. We follow the introductory paragraphs with the authors' informative abstract. Now draw upon the introductory paragraphs to devise a sentence or sentences to add to their abstract that might make it a more persuasive paragraph.

Introductory paragraphs

The construction of complex form from similar repeating units is a basic feature of spatial organization in all higher animals. Very little is known for any organism about the genes involved in the process. In *Drosophila* [fruit flies], the metameric [divided into similar segments] nature of the pattern is most obvious in the thoracic and abdominal segments of the larval epidermis [outer layer] and we are attempting to identify all loci required for the establishment of the pattern. The identification of these genes and the description of their phenotypes [visible characteristics] should lead to a better understanding of the general mechanisms for the formation of metameric patterns. . . .

We have undertaken a systematic search for mutations that affect the segmental pattern depending on the zygotic [a type of biological cell] genome. We describe here mutations at 15 loci which show one of three novel types of pattern alteration: pattern duplication in each segment (segment polarity mutants; six loci), pattern deletion in alternating segments (pair-rule mutants; six loci) and deletion of a group of adjacent segments (gap mutants; three loci). (Nüsslein-Volhard and Wieschaus 1980)

Abstract
In systematic searches for embryonic lethal mutants of *Drosophila melanogaster* we have identified 15 loci which when mutated alter the segmental patterns of the larva. These loci probably represent the majority of such genes in *Drosophila*. These phenotypes of the mutant embryos indicate that the process of segmentation involves at least three levels of spatial organization: the entire egg as developmental unit, a repeat unit with the length of two segments, and the individual segment. (Nüsslein-Volhard and Wieschaus 1980)

Answer

Revised abstract
The construction of complex forms from similar repeating units is a basic feature of spatial organization in all higher animals. Because very little is known for any organism about the genes involved in the process, we undertook a systematic search for embryonic lethal mutants of *Drosophila melanogaster*. As a result, we have identified 15 loci which when mutated alter the segmental patterns of the larva. These loci probably represent the majority of such genes in *Drosophila*. These phenotypes of the mutant embryos indicate that the process of segmentation involves at least three levels of spatial organization: the entire egg as developmental unit, a repeat unit with the length of two segments, and the individual segment.

CHECKLIST

If you are having trouble beginning a new abstract or revising an existing draft, we recommend that you run through the following checklist. We follow each question with an answer taken from an abstract that concerns a new computer model for simulating the evolution of a Universe permeated by a mysterious substance called "dark matter" (Spingel et al. 2005):

1. Ask yourself, what was done?
   "Here we present a simulation of the growth of dark matter structure . . . "
2. Ask yourself, how was it done?
   " . . . using $2{,}160^3$ particles, following them from redshift $z = 127$ to the present in a cube-shaped region 2,230 billion lightyears on a side. In post-processing, we also follow the formation and evolution of the galaxies and quasars . . . "

3. Ask yourself, what was discovered?
   "We show that baryon-induced features in the initial conditions of the Universe are reflected in distorted form in the low-redshift galaxy distribution, an effect that can be used to constrain the nature of dark energy with future generations of observational surveys of galaxies."
4. Decide whether you also want to indicate the significance of these results. To do so is to turn an informative abstract into its persuasive counterpart:
   "*The cold dark matter model has become the leading theoretical picture of the formation of structure in the Universe. This model, together with the theory of cosmic inflation, makes a clear prediction for the initial conditions for structure formation and predicts that structures grow hierarchically through gravitational instability. Testing this model requires that the precise measurements delivered by galaxy surveys can be compared to robust and equally precise theoretical calculations.* Here we present a simulation of the growth of dark matter structure using $2{,}160^3$ particles, following them from redshift $z = 127$ to the present in a cube-shaped region 2,230 billion lightyears on a side. In postprocessing, we also follow the formation and evolution of the galaxies and quasars. We show that baryon-induced features in the initial conditions of the Universe are reflected in distorted form in the low-redshift galaxy distribution, an effect that can be used to constrain the nature of dark energy with future generations of observational surveys of galaxies."
5. When you have finished your abstract, ask yourself: will readers be able to follow my train of thought without consulting the main text? In our view, the greatest failing of published abstracts is that they make little sense until you have read the main text. Creating a heading abstract that stands on its own is especially important today because most appear on a single Web page (with the title and byline) accessible to anyone while the whole article is viewable only by journal subscribers.

Remember that the abstract can also serve as a barometer that measures how well the full article works. Read your abstract carefully after having completed your article. Do you believe another scientist not intimately familiar with your work will understand your main message with ease? If not, the problem might not lie only in the abstract but in the article itself.

3 Entitling Your Research

The Modern Scientific Title

Scientific titles have evolved over time. Compared with titles from earlier times, modern ones are much more specific and technical, stripped of anything personal or openly literary. A typical seventeenth-century title on the subject of biology is "An Account of the Nature and Differences of the Juices, More Particularly, of Our English Vegetables." In the famous contents page reproduced in figure 1, the main title to Isaac Newton's first ever scientific article is similarly general and free of a technical vocabulary: "New Theory about Light and Colors." In contrast, the following recent title headed an important experimental letter in *Nature* written by a research group led by Lene Vestergaard Hau (Liu et al. 2001):

Observation of Coherent Optical Information Storage in
an Atomic Medium Using Halted Light Pulses

This typical modern title begins with the noun *observation*, implying the action "we experimentally observed." This noun is followed by the observed process, expressed in scientific English: "coherent optical information storage." The authors also include the location where the storage occurred ("atomic medium") and the means by which they achieved it ("halted light pulses").

One could not imagine such an information-packed title headlining a newspaper or popular science account. For comparison, the title for the *Scientific American* article reporting on this discovery is the poetic phrase "Frozen Light," while the *New York Times* goes for the more dynamic "Scientists Bring Light to Full Stop, Hold It, Then Send It on Its Way."

Let's now quickly run through select exemplary titles to classic *Nature* articles reproduced in the collection *A Century of Nature: Twenty-one Discoveries That Changed Science and the World* (Garwin and Lincoln 2003). For comparison, in parentheses, we include the titles to the specially commissioned explanatory commentaries that precede the reproduced articles. These com-

PHILOSOPHICAL
TRANSACTIONS.

February 19. 16$\frac{71}{72}$.

The CONTENTS.

A Letter of Mr.Isaac Newton,*Mathematick Professor in the University of Cambridge; containing his New Theory about* Light *and* Colors: *Where* Light *is declared to be not Similar or Homogeneal, but consisting of difform rays, some of which are more refrangible than others: And* Colors *are affirm'd to be not Qualifications of Light, deriv'd from Refractions of natural Bodies, (as 'tis generally believed;) but Original and Connate properties, which in divers rays are divers: Where several Observations and Experiments are alledged to prove the said Theory. An Accompt of some Books:* I. *A Description of the EAST-INDIAN COASTS, MALABAR, COROMANDEL, CEYLON,&c. in* Dutch, *by* Phil.Baldæus. II. Antonii le Grand *INSTITUTIO PHILOSOPHIÆ,*secundùm principia Renati Des-Cartes; *novâ methodo adornata & explicata.* III. *An Essay to the Advancement of MUSICK; by* Thomas Salmon *M.A. Advertisement about* Thæon Smyrnæus. *An* Index *for the Tracts of the Year* 1671.

FIGURE 1. Contents page from 1672 Philosophical Transactions (Newton 1672).

parisons illustrate the sharp difference between titles aimed at specialized and general audiences—a subject we will return to in chapter 11.

1. A Jupiter-Mass COMPANION to a Solar-Type Star
   (Seeking other solar systems)
2. Viable OFFSPRING Derived from Fetal and Adult Mammalian Cells
   (Dolly!)
3. A Three-Dimensional MODEL of the Myoglobin Molecule Obtained by X-ray Analysis
   (Dawn of structural biology)
4. Single Channel CURRENTS Recorded from Membrane of Denervated Frog Muscle Fibres
   (Molecular switches for "animal electricity")
5. 3C 273: A Star-like OBJECT with Large Red-Shift
   (The quasar enigma)
6. The SCATTERING of Electrons by a Single Crystal of Nickel
   (Electrons make waves)

These scientific titles succinctly capture the authors' major claims within a short phrase; they do so by having as their nucleus a noun that specifies the main object or concept or action behind the discovery (emphasized in small caps in the examples), then elaborating with modifiers fore and aft. You cannot delete a single term without fatally damaging the original.

The alert reader will have noted that the fifth title is slightly different from the others in that it has two nucleus nouns: one in the main title, the other in the subtitle. The main title is simply the name assigned to the stellar object under scrutiny; the subtitle explains what about *3C 273* makes it worthy of other astronomers' attention. This subtitle follows the same pattern as the other four titles: nucleus noun with explanatory modifiers to the left and right.

What are modern scientific titles for? To answer that, we must first know why individuals read scientific articles in the first place. First and foremost, scientists read articles to help formulate new research problems or apply the content to their present research. Secondary reasons include staying abreast of the latest developments in their field and incorporating these into lectures to students. This means that a successful title works together with the abstract and introduction to urge potential readers that reading on will be of value to them professionally.

Titles serve another critical purpose. A key part of preparing to write a research proposal or scientific article is searching the literature to check whether anyone else has already solved the same or similar problem. That normally involves a computer search of existing titles. Confusing, misleading, or weak titles increase the likelihood that the searcher will miss relevant published literature. And that can lead to a waste of time and research funds in a needless duplication of effort.

Claim-Staking Titles

We think of titles as coming in three varieties. The first and most common captures the major claim; the second poses the problem to be solved; the third encapsulates the theme of the article. Appended to any one of these types might be a few words concerning the method used. If you have read chapters 1 and 2, you ought to recognize these formulations; they are essentially the same elements that go into an introduction and abstract. Thus the title, abstract, and introduction in a scientific article all work together to serve the same communicative end—informing readers up front about a new solution to a research problem.

We now take a closer look at the claim-staking title "Observation of Coherent Optical Information Storage in an Atomic Medium Using Halted Light

Pulses." This title captures the authors' main claim to new knowledge: they have discovered a way to store coherent optical information. It also incorporates the method: the slowing of light pulses to a complete stop in an atomic medium. This title is a concatenation of a nominalized action (observation, in italics) and three strings of key technical terms (underscored), all fused into a single phrase:

Observation
of Coherent Optical Information Storage
in an Atomic Medium
Using Halted Light Pulses

How might the Hau research group have arrived at their title? The first three elements appear in a key sentence from the authors' conclusion: "We have *demonstrated experimentally* [observation] that *coherent optical information* can be *stored* [coherent optical information storage] in *an atomic medium* [atomic medium] and subsequently read out by using the effect of EIT in a magnetically trapped, cooled atom cloud" (our emphasis). And the last technical term appears as a short phrase in the key sentence of their first paragraph: "Here we use electromagnetically induced transparency *to bring laser pulses to a complete stop* [halted light pulses] in a magnetically trapped, cold cloud of sodium atoms" (our emphasis). Scientific titles are often pared-down versions of the solution statements in the abstract, introduction, or conclusion sections.

Grammatically, claim-staking titles work by putting in place what we have called a "nucleus" and what linguists call a "head noun." Modifying words or phrases are then placed to the right and sometimes the left of that word. For this particular title, the head noun is *observation,* and the words to the right act in a support role:

OBSERVATION [*nucleus noun*] → of Coherent Optical Information Storage in an Atomic Medium Using Halted Light Pulses

Claim-staking titles work best when the nucleus noun expresses the essence of the central discovery. This principle reveals a possible shortcoming with the *Nature* title just quoted. The nucleus noun is *observation.* But what does it mean to say that the authors "observed" information storage? Does that word add much? The key to this paper is really the storage of optical information, not its observation. The authors and *Nature* editors might, we realize, take issue with our reading. Still, for the sake of argument, let's change the nucleus noun from *observation* to *storage* simply by deleting "observation of." Here's the result:

Coherent Optical Information ← STORAGE [*new nucleus noun*] →
in an Atomic Medium Using Halted Light Pulses

The emphasis now lies on the nucleus noun *storage*. Next, let's shift the new head noun to where it will stand out more—in the prominent first position:

STORAGE → of Coherent Optical Information
in an Atomic Medium Using Halted Light Pulses

Suppose that instead of its storage we wanted to emphasize the truly startling achievement of bringing light pulses to a stop in an atomic medium of supercold sodium. Then we would rearrange the title as follows:

HALTING [*new nucleus noun*] → of Light Pulses for Storage of
Coherent Optical Information in an Atomic Medium

This third alternative has the advantage of a cause-effect structure: the "halting of light" is the cause, the "storage of information" the effect.

In both this third revision and the original title, the nucleus noun names an action: *to halt* in the former, *to observe* in the latter. We prefer the former because it refers to an action central to the major claim being made. But whatever version of the *Nature* title one prefers, the key point here is that the nucleus noun dominates and controls the title meaning.

Let's now look more closely at the nucleus noun itself. Our next title is from one of the most famous papers in all the twentieth-century scientific literature, the 1953 paper in *Nature* by James Watson and Francis Crick:

A STRUCTURE for Deoxyribose Nucleic Acid

No confusion or extra words there. Encapsulated in that short title is a straightforward statement of Watson and Crick's claim that they uncovered the "structure" of DNA. As you can deduce from the nucleus noun, the emphasis is not on DNA, which they did not discover, but its structure, which they did. The authors could have expressed their main claim more fully and forcefully by adding a few modifying words:

The Double-Helical Structure of Deoxyribose Nucleic Acid

or even

The Double-Helical Structure of the Genetic Molecule DNA

Their choice is a more compact and more tentative. "A Structure . . . " (rather than "The Structure . . . ") leaves open the possibility that other structures could exist or even that theirs could be mistaken. That is the message

Watson and Crick felt most comfortable with at the time. The two authors did not mention genetics because they were preparing a separate paper on that topic: "Genetical Implications of the Structure of Deoxyribonucleic Acid." Not all factors that go into decisions about titling have to do with clear and concise expression or capturing the main claim in a short phrase.

Problem-Posing Titles

Yet another style of title comes in the form of a question that encapsulates the problem the article addresses. Albert Einstein was fond of this title style:

Does the Inertia of a body Depend on Its Energy Content? (1905a)

Can Quantum-Mechanical Description

Really Be Considered Complete? (1935)

Here is another tantalizing title posed as a geological question related to the role of sea-floor spreading in plate tectonics:

Did the Atlantic Close and Then Re-open? (Wilson 1966)

This title style has the advantage of appealing directly to the curiosity of the potential reader. We do not find that many scientific titles in fact conform to this style. Yet most claim-staking titles can be easily recast into problem-posing ones. We do so by turning several claim-staking titles just discussed into questions:

What Is the Structure of Deoxyribose Nucleic Acid?

What Is the Structure of the Myoglobin Molecule?

Can Coherent Optical Information Be Stored in an Atomic Medium?

Can Viable Offspring Be Derived by Cloning
Fetal and Adult Mammalian Cells?

Is There a Jupiter-like Planet Orbiting a Solar-Type Star?

A check on whether you have indeed written a good title or not is to rewrite it as a question. If that question concisely captures your main research problem, then you have succeeded.

Thematic Titles

Another common title strategy is simply to introduce the main theme. To illustrate a thematic title, we have chosen a famous review article written by the French biologists François Jacob and Jacques Monod (1961). It proposes a daring theoretical model for how protein gets made from DNA. James Watson captured this process in the aphorism "DNA makes RNA makes protein." Of course, the actual process is a little more complicated than that, as is immediately evident from Jacob and Monod's abstract:

> The synthesis of enzymes in bacteria follows a double genetic control. The so-called structural genes determine the molecular organization of the proteins. Other, functionally specialized, genetic determinants, called regulator and operator genes, control the rate of protein synthesis through the intermediacy of cytoplasmic components or repressors. The repressors can be either inactivated (induction) or activated (repression) by certain specific metabolites. This system of regulation appears to operate directly at the level of the synthesis by the gene of a short-lived intermediate, or messenger, which becomes associated with the ribosomes where protein synthesis takes place.

It is hard to image a reasonably compact claim-staking title that could do justice to the gist of that complex paragraph. The authors did not try. Instead, they just stated their topic: "Genetic Regulatory Mechanisms in the Synthesis of Proteins."

Structurally, this title works in the same way as the claim-staking ones: nucleus noun with modifying words before and after. We learn from the title that this article centers on newly discovered "mechanisms" that control protein synthesis, and the abstract bears out that inference.

Other Title Styles

Another title style joins two nouns, implying a hitherto undiscovered relationship between them. An example is "Super-novae and Cosmic Rays," from a 1934 *Physical Review* article by Walter Baade and Fritz Zwicky. These two men are forever linked by their bold theoretical calculations indicating that "cosmic rays are produced by super-novae [exploding stars]." Structurally, the Baade-Zwicky title differs from the earlier ones we quoted. It has no single nucleus noun; rather it joins two nucleus nouns with a conjunction. You might wonder why not the reverse order, "Cosmic Rays and Super-novae." It is because at bottom, this article seeks to answer the question "What are super-novas?" not "What are cosmic rays?" In any hierarchical list of two or more items, you should consider ordering them according to degree of importance.

On occasion, authors of scientific papers solicit the reader's attention with an unorthodox title. P. J. E. Peebles and Joseph Silk's "A Cosmic Book of Phenomena" from 1990 *Nature* is one such example. This lighthearted vein carries over into the opening paragraph:

> There is a tendency at scientific meetings, when a particularly important but tentative result is presented, to demand of one's colleagues what odds they would give for eventual confirmation. Many bottles of the finest champagnes and malt whiskies, and even more esoteric stakes, rest in abeyance while observers struggle to count rare photons from remote galaxies or experimentalists devote decades to designing new types of detectors. To enrich, enlighten and even amuse those of our colleagues who are trying to assess the merits of the rival cosmogonies, we have begun a modest programme of setting up a cosmic book of odds. Our first book focused almost exclusively on the large-scale structure of the Universe. This [present] one is devoted to the observable phenomena that theorists customarily invoke (or ignore) in developing models for the formation of the galaxies.

Note than despite the somewhat unorthodox title, its nucleus noun *book* does refer to the main outcome of the authors' research, the "cosmic book of odds" referred to in the above paragraph. In general, your title's nucleus noun should figure prominently in the solution component of the introduction (chapter 1).

Building a More Informative Title

Let's illustrate our principles with yet a more extended example. Consider the following title (Cacace et al. 1996):

> The Gas-Phase Reaction of Nitronium Ion with Ethylene:
> Experimental and Theoretical Study

We believe that neither this title nor its subtitle gives us complete sense of the authors' achievement. After reading the abstract, however, we can revise appropriately.

> ABSTRACT. The addition of NO_2^+ to ethylene, the prototypical electrophilic nitration of a π system and the focus of considerable theoretical interest as a model of aromatic nitration, has been studied in the gas phase by FT-ICR, MIKE, and CAD mass spectrometry, complemented by ab initio calculations at the MP/26–31+G* level of theory. The results provide a clearcut answer to the principal mechanistic question addressed, showing that the reaction yields a O-nitroso product, probably CH_3CHONO^+, rather than a C-nitrated product.

First suggested revision of title
O-Nitroso: The Product of Gas-Phase Aromatic Nitration

Second suggested revision of title
The Gas-Phase Reaction of Nitronium Ion with Ethylene: A New Model for Aromatic Nitration

Third suggested revision of title
A New Model for Aromatic Nitration: O-Nitroso Formed by Reaction of Nitronium Ion with Ethylene

From the abstract, what would you say is the most important point or points the authors make? We find two. First and most important, the authors formed something called "O-nitroso" by a chemical reaction. Second, this experimental achievement is of "considerable theoretical interest" in understanding a chemical process called "aromatic nitration." Wouldn't the reader be better clued in to the paper's contents if at the very least "O-nitroso" appeared in the title? In the first suggested revision, we give it pride of place as the head noun: "O-Nitroso: The Product of Gas-Phase Aromatic Nitration."

Obviously, the second important point ("theoretical interest") did not make it into our first revision. To that end, we turn to the subtitle, "Experimental and Theoretical Study," the phrase after the colon in the original. Presumably the authors appended it to emphasize that their study involved not only collecting experimental data on the gas-phase reaction but also determining the same property by means of a theoretical model and then comparing the two sets of figures. If readers had to rely exclusively on the original subtitle, without access to the abstract, we question whether they would get that message. We think that the authors could have used that space more meaningfully:

Second suggested revision
The Gas-Phase Reaction of Nitronium Ion with Ethylene: A New Model for Aromatic Nitration

The advantage of this second revision is that we show our chemist-readers why they should care that we studied this particular reaction in the lab.

If, on the other hand, the authors believe the theoretical part more important than the experimental, then they could reverse this order and revise the subtitle:

Third suggested revision
A New Model for Aromatic Nitration: O-Nitroso Formed by Reaction of Nitronium Ion with Ethylene

In the second revision the emphasis is on the chemical reaction, in the third on the new model.

We close with a caveat: the original title is not only accurate but, in fact, good enough for publication in a prestigious scientific journal. Still, we believe that the suggested revisions are more informative. By analyzing your title as we have just done, you can maximize your chances that potential readers will become actual readers.

EXERCISES

A good title is typically an abstract of the abstract. For practice, try concocting informative titles from the following three abstracts. Bear in mind that, in contrast to a math problem, these exercises have no one right answer.

1. The human genome holds an extraordinary trove of information about human development, physiology, medicine and evolution. Here we report the results of an international collaboration to produce and make freely available a draft sequence of the human genome. We also present an initial analysis of the data, describing some insights that can be gleaned from the sequence. (International Human Genome Sequencing Consortium 2001)
2. A DNA sequence for the genome of bacteriophage ΦX_{174} of approximately 5,375 nucleotides has been determined using the rapid and simple "plus and minus" method. The sequence identifies many of the features responsible for the production of the proteins of the nine known genes of the organism, including initiation and termination sites for the proteins and RNAs. Two pairs of genes are coded by the same region of DNA using different reading frames. (Sanger et al. 1977)
3. During experiments aimed at understanding the mechanisms by which long-chain carbon molecules are formed in interstellar and circumstellar shells, graphite has been vaporized by laser irradiation, producing a remarkably stable cluster consisting of 60 carbon atoms. Concerning the question of what kind of 60-carbon atom structure might give rise to a superstable species, we suggest a truncated icosahedron. A polygon with 60 vertices and 32 faces, 12 of which are pentagonal and 20 hexagonal . . . The C_{60} molecule that results when a carbon atom is placed at each vertex of this structure has all the valences satisfied by two single bonds and one double bond, has many resonance structures, and appears to be aromatic. (Kroto et al. 1985)

Answers

1. For the first abstract, we judged the last two sentences to be key to the major claim, while the first sentence concerns the importance of the claim to society. Our own guess at a scientific title was

   Initial Analysis of Data from Sequencing the Human Genome

The original title is similar but has a slightly different emphasis:

Initial Sequencing and Analysis of the Human Genome

The head noun for our version is "analysis" only, while that of the original is the compound "sequencing and analysis," which better reflects the contents of the whole article.

Of course, neither of those titles would do for a lay audience. Can you create a more literary title from the abstract's first sentence? How about "Unlocking the Treasure of the Human Genome"?

2. The original correctly stresses the sequencing aspect of the discovery:

   Nucleotide Sequence of Bacteriophage ΦX_{174} DNA

   Had they chosen to be a little more detailed, the authors might have added to the right of the head noun a little more information about the organism selected for sequencing:

   Nucleotide Sequence Responsible for All Nine Genes of Bacteriophage ΦX_{174}

3. The original title is short and catchy:

   C_{60}: Buckminsterfullerene

   That title emphasizes the size of the newly discovered carbon molecule (60 atoms) along with the authors' new name for it. You would not have guessed that title, because the abstract does not mention the new carbon's name. You may have proposed a title along the lines of the following, emphasizing the molecule's size, special property, and structure:

   C_{60}: A Truncated Icosahedral Structure for a Superstable Form of Carbon

CHECKLIST

We close with a checklist you can consult while constructing a new or revising an existing claim-staking title, the most popular type:

1. Does your title accurately reflect your main discovery?
2. Is the gist of your discovery anchored in a nucleus noun?
3. Is that nucleus noun placed most prominently in your title?
4. Do the words and phrases that supplement your nucleus noun refer to important aspects of your discovery, aspects you want no one to miss?
5. Can your title stand alone as an independent encapsulation of the central content of your article, in effect an abstract of your abstract?

4 Turning Your Evidence into Arguments

RESULTS AND DISCUSSION

We now say goodbye to the front matter and hello to the heart of the matter: results and their discussion. In modern articles we find much variation in how authors divide their material between these sections. Sometimes results appear separately from discussion; at other times the two are combined. Long articles tend toward the former organization, shorter ones toward the latter. Moreover, as John Swales (1990) observes, there is "much variation in the extent to which Results sections simply describe results and the extent to which Discussion sections redescribe results." In other words, the results section emphasizes results but contains some discussion, while the discussion section emphasizes discussion but restates some results.

For the results and discussion we cannot articulate a typical structure as we did for the introduction and abstract. None exists as far as we can discern. But we can discuss some common threads. In results, our emphasis will be on how to (1) present your findings in tables, figures, and their accompanying text and (2) show how the limitations of your methods qualify the factual status of your findings. In discussion, our emphasis will be on how to (1) construct an argument that turns the findings from results into evidence for new scientific claims and (2) limit and qualify your claims so that they are in conformity with the evidence.

Presenting Results

See if you can make any sense out of the next paragraph of results, composed by us by drawing upon the contents of a famous astronomical paper from 1929:

> We have estimated the radial velocity (v) and distance (r) relative to the earth for 24 nebulae. The distances were estimated from the apparent luminosity of the brightest stars in the nebulae or from the mean luminosities in a cluster. The radial velocities, corrected for solar motion, were determined from the red-shifts measured at the Mount Wilson Observatory. The results are as follows: S. Mag., v = 170 km/sec, r = 0.032 $\times$ 10^6 parsecs; L. Mag.,

> v = 290 km/sec, r = 0.034 x10^{6} parsecs; N.G.C. 6822, v = 130 km/sec; r = 0.214×10^6 parsecs; 598, v = 70 km/sec, r = 0.263×10^6 parsecs; 221, v = 185 km/sec, r = 0.275×10^6 parsecs; 224, v = 185 km/sec, r = 0.275×10^6 parsecs; 5457, v = 200 km/sec, r = 0.45×10^6 parsecs; 4736, v = 290 km/sec, r = 0.5×10^6 parsecs; 5194, v = 270 km/sec, r = 0.5×10^6 parsecs; 4449, v = 200 km/sec, r = 0.275×10^6 parsecs; 4214, v = 300 km/sec, r = 0.8×10^6 parsecs; 3031, v = 30 km/sec, r = 0.9×10^6 parsecs; 3627, v = 650 km/sec, r = 0.9×10^6 parsecs; 4826, v = 150 km/sec, r = 0.9×10^6 parsecs; 5236, v = 500 km/sec, r = 0.9×10^6 parsecs; 1068, v = 920 km/sec, r = 1.0×10^6 parsecs; 5055, v = 450 km/sec, r = 1.1×10^6 parsecs; 7331, v = 500 km/sec, r = 1.1×10^6 parsecs; 4258, v = 500 km/sec, r = 1.4×10^6 parsecs; 4151, v = 960 km/sec, r = 1.7×10^6 parsecs; 4382, v = 500 km/sec, r = 2.0×10^6 parsecs; 4472, v = 850 km/sec, r = 2.0×10^6 parsecs; 4486, v = 800 km/sec, r = 2.0×10^6 parsecs; 4649, v = 1090 km/sec, r = 2.0×10^6 parsecs.

Be honest. Did you even read that entire final sentence? If you did, we greatly admire your tenacity, but we guess that nearly all readers will skim that thicket of quantitative information, if they look at it at all. Yet within that final sentence lies one of the profound secrets of the cosmos. As expressed in the above passage, however, few if any readers would ever guess it. Now, let's convert that sentence into a table of data—see table 1.

We extracted the three columns of table 1 from a larger table that appeared in a 1929 article by the early twentieth-century master of astronomy Edwin Hubble. Compare the last sentence in our made-up passage with the table. In tables, exploitation of the vertical and horizontal dimensions of the page minimizes the effort involved in finding and identifying data elements; it makes comparisons among these elements easy; it turns a sequence of comparisons into a single unified operation that suggests trends, the first step in untangling significance.

An analysis of our streamlined version of Hubble's table reveals that the first column lists galaxies (referred to by Hubble as "extra-galactic nebulae") by their numeric designations. Their order is dictated by the second column, the estimated distance of the galaxy from Earth. As we move down the second column, the distances from Earth steadily increase for the two sets of galaxies, the first pair, and then the remaining twenty-two. The third column lists the velocities determined by a then newly invented technique, measuring shifts in the light spectrum ("red-shift") created as any luminous celestial body speeds away from planet Earth.

Can you spot trends in those data, even when presented in an orderly tabular fashion? Perhaps, but for that a graph works better: a graph allows

TABLE 1. Distances and velocities for extra-galactic nebulae known in 1920s. Adapted from Hubble 1929.

Galaxy	Distance from Earth (million parsecs)	Velocity (km/sec)
S. Mag.	0.032	+170
L. Mag.	0.034	+290
NGC 6822	0.214	−130
598	0.263	−70
221	0.275	−185
224	0.275	−220
5457	0.45	+200
4736	0.5	+290
5194	0.5	+270
4449	0.63	+200
4214	0.8	+300
3031	0.9	−30
3627	0.9	+650
4826	0.9	+150
5236	0.9	+500
1068	1.0	+920
5055	1.1	+450
7331	1.1	+500
4258	1.4	+500
4151	1.7	+960
4382	2.0	+500
4472	2.0	+850
4486	2.0	+800
4649	2.0	+1090

NOTE: A parsec (parallax of one arc second) is an astronomical unit equivalent to 3.3 light years (19 trillion miles). "S. Mag." and "L. Mag." stand for small and large magellanic cloud, respectively. And NGC stands for the *New General Catalogue,* a comprehensive astronomical catalog first compiled in the 1880s.

you to see unequivocally what a table permits you to infer only laboriously and tentatively. The graph that Hubble derived from his table is presented here as figure 2.

Because of the scatter among the data points, Hubble drew two lines of central tendency as a guide to the eye—two readings of the displayed data, both with the same message. The solid line represents the central tendency of the twenty-four galaxies listed in the table (solid circles), and the dotted line represents the central tendency of the same galaxies combined in nine groups "according to proximity in direction and in distance" (open circles). The cross represents the mean velocity corresponding to the mean distance of twenty-two additional galaxies not listed in the table because their individual distances could not be estimated reliably. While "the data in the table indicate a linear correlation between distances and velocities," the graph makes that relationship explicit at a glance.

Visuals like Hubble's do not normally stand by themselves. Full understanding requires explanatory text in captions and the results or discussion section. Readers need definitions of graphical elements such as data symbols, lines and curves, and any unfamiliar terms or abbreviations in legends or on the axes. Readers also need interpretation of data trends: Do the data increase, decrease, vary irregularly, vary because of some change in test conditions, or remain constant? Do the data appear to be randomly distributed or change in agreement with a mathematical formula? Do the data contradict or add to current understanding? What's more, readers need to learn about any qualifications to the data interpretation. The motto of the first major scientific society, the Royal Society of London, was "Trust no one's word" (in Latin, *Nullius in verba*). That motto remains apt today in a somewhat different sense from that of the seventeenth century, when experimentation under controlled

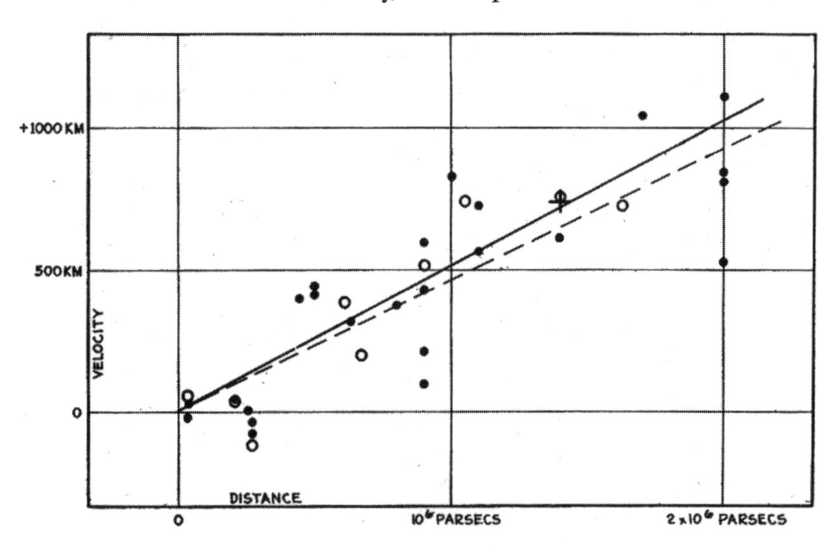

Velocity distance relation among extra-galactic nebulae.

FIGURE 2. Visual display of data from table 1 in graph form after correction of velocities for solar motion (Hubble 1929).

conditions was still in its infancy and statistical analysis was virtually nonexistent. In today's scientific world, the injunction might be interpreted to mean "Demonstrate that you have considered all possible sources of error in gathering your data." To do so, you must resort to text. Here is what Hubble (1929) wrote about the reliability of the distance results recorded in his table and graph:

> The first seven distances are the most reliable, depending . . . upon extensive investigations of many stars involved. The next thirteen distances . . . are subject to considerable probable errors but are believed to be the most reasonable values at present available. The last four objects appear to be in the Virgo Cluster. The distance assigned to the cluster, 2×10^6 parsecs, is derived from the distribution of nebular luminosities, together with luminosities of stars in some of the later-type spirals, and differs somewhat from the Harvard estimate of ten million light years [3×10^6 parsecs].

Qualifying your results as Hubble does above is not merely a rhetorical nicety; it is your responsibility as an author. If you do not, someone else will likely do so in less flattering terms. The astronomers responsible for the "Harvard estimate" might have thought Hubble incompetent or delusional had he not spelled out his reasoning for the last four distances in his table. They still probably questioned his estimate, but at least they were told why he decided upon 2×10^6 parsecs instead of 3×10^6. The Harvard estimate aside, most knowledgeable readers at the time would not have challenged the accuracy of Hubble's distance estimates as a whole but would have acceded to his explanation that they were "the most reasonable values at present available."

When you are deciding how to best present your results, we suggest you consider the following:

- Tables have an important advantage in that they record exact values and facilitate the comparison of series of data. The original Hubble table has six columns. Incorporating all that information into a single graph would have been impossible, or at least impossibly messy. The main shortcoming of tables is that they are not particularly useful for discovering or communicating data trends.
- Graphs have the considerable advantage that the relationship among several data sets can be viewed at a glance and trends that are not readily evident in columns of data can be conveyed. Graphs have the disadvantage that the viewer cannot as a general rule determine exact values. There are also limits to the number of points and curves you can cram into a single graph without having it deteriorate into a tangle—the visual equivalent

of the endless and incomprehensible sentence we derived from Hubble's table. At the other end of the spectrum, when relatively few data points are involved, tables or descriptive text tends to work best.

- Text allows you to explain the results shown in your tables or figures. One of the shortcomings of many results and discussion sections is that the authors leave readers to their own devices in interpreting the full meaning of tables and figures. Tables and figures are not like pictures in a museum exhibition: you cannot leave meaning up to viewers.

Discussing Results: Comparisons

Arguments made about results often center on comparisons in various forms. In analyzing and presenting your results you should always bear in mind the question "Compared to what?" The following sentence, for example, does not tell us much about the important scientific achievement that it reports:

> In 1986 a ceramic made of Ba-La-Cu-O was found to become superconducting at 35 kelvin [degrees above absolute zero].

Appending a simple comparative begins to overcome this deficiency:

> In 1986 a ceramic made of Ba-La-Cu-O was found to become superconducting at 35 kelvin, more than 10 degrees above the previous high.

But readers might still ask: What is the significance of the 10-degree rise in superconducting temperature? Such readers need to know that the temperature of a relatively cheap and readily available coolant, liquid nitrogen, is 77 kelvin. In addition, these readers need to know that researchers had previously thought the upper superconductivity limit was 20–25 kelvin. A sizable increase in that upper limit suggested a new possibility: the development of a cable cooled by liquid nitrogen that offered no resistance to the passage of electricity, a feat once thought impossible.

It is by such comparisons that authors make sense of their data for themselves and us. Comparisons tend to permeate discussion sections: outcomes from an experimental group versus a control group, measurements of some physical property at an initial state versus an altered state, present results versus those reported previously by others, experimental measurements versus theoretical calculations, measurements or calculations for the same property obtained by different methods, and so on.

A real-world example of argument from comparison is Hubble's. In his 1929 article on the velocity of faraway celestial objects, Hubble determines distances by comparing luminosities, compares distances to velocities to determine a trend, and compares the trend to the existing theory (originated by Willem de Sitter) that the mean density of matter in the universe is such that it will neither expand indefinitely nor collapse.

Another example is the article on child psychology by Bandura, Ross, and Ross (1961). They compare the behavior of children who had witnessed aggressive adult actions with those who did not. In one trial, the experimental group saw an adult aggressively knock a doll off a container, then some time later replace it. The control group saw the adult "accidentally" knock the doll off the same container, but the adult did not replace it. Here is the explanation of what the two groups did when they subsequently saw an adult knock the doll off the container without replacement: "While the control subjects replaced the doll on the box slightly more often than the subjects in the experimental group, this difference tested by means of the median test was not statistically significant. . . . Evidently the response of replacing things, undoubtedly overtrained by parents, is so well established that it occurs independently of the behavior of the model." The authors compared experimental and control groups and found a difference, but they concluded that it was not statistically significant: this finding did not support their main hypothesis that children learn by imitating adults. As a consequence of this failure, the authors felt a need to include a justification for considering this situation an exception. This sort of verbal analysis of the acquired evidence is typical of what one finds in good discussion sections.

In most discussion sections, authors elaborate on their achievement by comparing it with work done earlier by others. As an example, we quote from the discussion section reporting the development of a DNA-based computer that plays tic-tac-toe against a human opponent and never loses. The authors named their molecular automaton MAYA (Stojanovic and Stefanovic 2003):

> It is instructive to compare MAYA with the other published molecular approaches[2–5,24–28] to computation in solution using DNA. MAYA does not take advantage of massively parallel computation, used in the Adleman-Lipton paradigm,[4] nor does MAYA use the power of DNA to form well-defined supramolecular complexes, as in the Seeman-Winfreee paradigm.[24,28] MAYA is unique in that it consists of individual bimolecular building blocks that behave analogously to digital logic circuits used in electronic computers or to enzymes in metabolic circuits.

In this passage, Stojanovic and Stefanovic differentiate their DNA computer from other "molecular approaches" and argue for its uniqueness and importance by comparison with similar ones made by others.

Comparisons need not involve data or references. Figure 3 is from an ex-

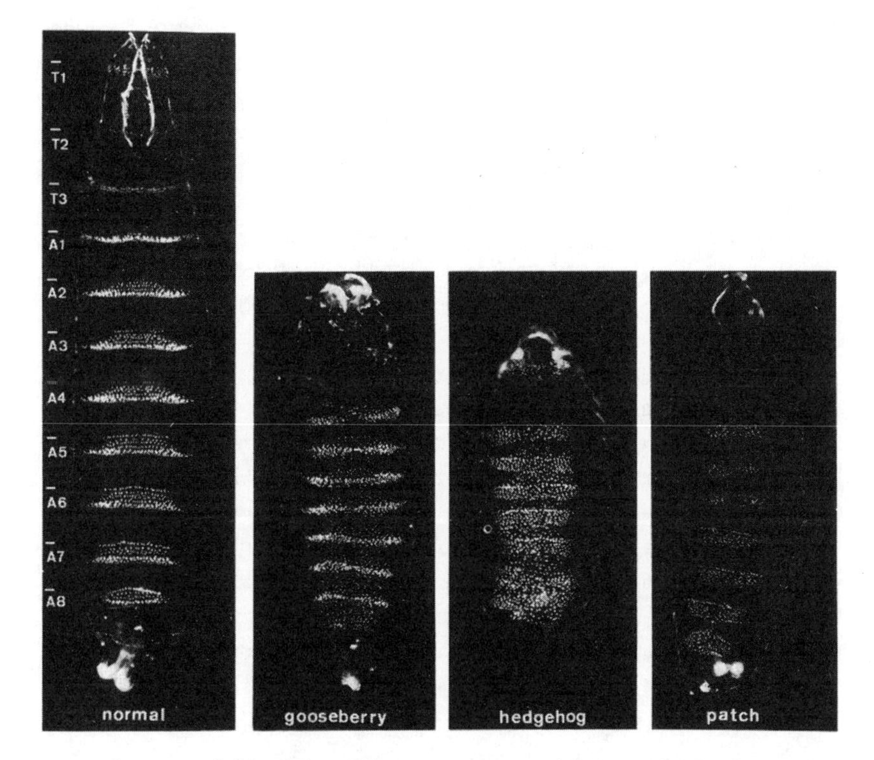

Ventral pattern of (from left to right) a normal *Drosophila* larva shortly after hatching, and larvae homozygous for *gooseberry*, *hedgehog*, and *patch*. The mutant larvae were taken out of the egg case before fixation. All larvae were fixed, cleared and mounted as described in ref. 22. A, abdominal segment; T, thoracic segment. X 140

FIGURE 3. Visual display comparing photographs of normal and mutant fruit-fly larvae. Reprinted with permission of Macmillan Publisher Ltd. (Nüsslein-Volhard and Wieschaus 1980).

perimental study by German scientists Christiane Nüsslein-Volhard and Eric Wieschaus (1980) on the processes that govern the embryonic development of body segmentation in fruit flies, one of the principal organisms used in the study of genetics. This figure compares normal and mutant larvae shortly after hatching. Its dramatic impact comes from direct visual comparison between the normal and the abnormal. The normal fruit-fly larva has an elongated oval shape with three thoracic and eight abdominal segments (far left photograph, marked T1 to T3 and A1 to A8, respectively). One end eventually develops a head, the other a tail. The authors' experiments revealed that, out of the many thousands of fruit fly genes, only about fifteen mutants controlled the embryonic development of these patterns of segmentation. In this work, the authors assign these mutants whimsical names like "gooseberry," "hedgehog," and "patch."

Nüsslein-Volhard and Wieschaus divided the fifteen mutants into three groups based on comparisons with the norm. Their discussion of results describes the normal segmental pattern, then each of the three mutant groups in turn. The order of presentation is driven by the type of deviation from the norm: changes in each segment, every other segment, and a connected series of segments. This important work established for the first time that researchers can identify the genes controlling embryonic development by such visual comparisons of larvae.

The above may make it seem as though making sound comparisons in scientific research is as easy as tic-tac-toe. It often is not. Return to the Hubble astronomical graph of velocity versus distance. Remove the solid and dashed lines, and you will clearly see wide scatter in the data. For that reason, Hubble characterized the linear velocity-distance relationship as "*a first approximation* representing a restricted range in distance" (our emphasis). Error can easily creep in as part of either research design or data collection and analysis. Providing practical advice on such technical matters is far beyond the scope of our book. However, we recommend that your discussion of any comparison should take note of any methodological issues that might qualify confidence in the results—the subject of our next section.

Discussing Results: Qualifications

The discussion section ought normally reflect what would in real life be an abnormal condition: a split personality. One personality boldly makes claims about the nature and causal structure of the material world; the other tempers those claims in an attempt to avoid misleading the community of scientists about the degree of certainty endorsed. As Wayne Booth, Gregory Colomb, and Joseph Williams advise in *The Craft of Research* (1995), "Though it may seem paradoxical, your argument gains rhetorical strength when you acknowledge its limits."

In the discussion section, then, scientists routinely interpret their results so as to make the broadest claims they plausibly can, given their evidential base. Nevertheless, they must at the same time be extremely cautious about making broad claims, lest their work be dismissed as rank speculation. Watson and Crick, bear in mind, spoke of *a*, not of *the*, structure of DNA in the title of their famous *Nature* article. This judgment does not reflect their subjective certainty, which, as we know from Watson's 1968 memoir, *The Double Helix*, was high to the point of exhilaration. It reflects rather a sense of professional humility in the face of the undoubted fact that while much science turns out to be right and deservedly takes its place within the permanent storehouse of knowledge, many promising knowledge claims are eventually seriously revised or discarded.

Nüsslein-Volhard and Wieschaus fully exhibit that combination of daring and caution characteristic of the split scientific personality. At the end of their article, they speculate boldly about the character of the processing mechanism whereby segmentation takes place; their support of their own claims, however, is suitably nuanced, an effect they achieve by the use of so-called hedging expressions, which we have italicized for emphasis:

> It is also *possible* that the double segmental units are never defined by distinct borders in normal development. The existence of a double segmental homology unit *may merely reflect* a continuous property such as a wave with a double segmental period responsible for correct spacing of segmental boundaries. . . . We have not found any mutations showing a repeat unit larger than two segments. This *may indicate* that the subdivision of the blastoderm proceeds directly by the double segmental repeat with no larger intervening homology units. However, the failure to identify such larger units *may reflect* the incompleteness of our data.

Besides the careful hedging of statements that might otherwise be viewed as overly speculative, good discussion sections spell out any conditions under which a new knowledge claim might cease to be valid or address legitimate criticisms or reservations that a fellow expert might raise. To give you a sense of this form of qualification, we have chosen the final discussion paragraph of the classic experimental article by Oswald Avery, Colin MacLeod, and Maclyn McCarty (1944) arguing that the complex DNA molecule harbors a coded message with genetic information:

> It is, of course, possible that the biological activity of the substance described is not an inherent property of the nucleic acid but is due to minute amounts of some other substance adsorbed to it or so intimately associated with it as to escape detection. If, however, the biologically active substance isolated in highly purified form as the sodium salt of desoxyribonucleic acid actually proves to be the transforming principle, as the available evidence strongly suggests, then nucleic acids of this type must be regarded not merely as structurally important but as functionally active in determining the biochemical activities and specific characteristics of pneumococcal cells. Assuming that the sodium desoxyribonucleate and the active principle are one and the same substance, then the transformation described represents a change that is chemically induced and specifically directed

> by a known chemical compound. If the results of the present study of the chemical nature of the transforming principle are confirmed, then nucleic acids must be regarded as possessing biological specificity the chemical basis of which is as yet undetermined.

As a more recent example, we quote an article on the pathogenesis of Alzheimer's disease by W. Taylor Kimberly and coworkers (2003), who identified the four membrane proteins of the γ-secretase complex, a key enzyme in the disease. Their discussion admits that

> our data cannot rule out the requirement for another small member of the complex that is normally in the cells we evaluated. Addressing this issue will require purification of the γ-secretase complex to homogeneity with retention of protease activity and complete analysis of all the purified components.

The trick here is to put yourself in the shoes of a highly critical reviewer of your research and then to pinpoint any weaknesses in, or limitations to, your initial argument. Often, however, authors do not possess enough critical distance to recognize certain problems that may be readily apparent to knowledgeable readers. You can gain that distance by several means: presentation of your preliminary results and conclusions at a scientific meeting or seminar, review of an early draft by colleagues, review of your manuscript by journal reviewers, or some combination of these.

Conclusion

We have avoided a separate chapter on scientific tables and visuals because we believe that scientific communication represents a synergy of text, tables, and visuals, three modes whose integration constitutes scientific meaning. This integration implies a division of labor. Tables record data; graphs indicate data trends; text explains, interprets, and qualifies. The goal is always the support of a claim of new scientific knowledge; the vehicle for this claim is nearly always an argument at whose heart is comparison—within data sets, between experimental and control groups, between experimental results and theories, between present and previously published results.

There is also a division of labor between results and discussion. Results generally focus on the generation of data—on, if you will, the facts and the most immediate inferences derivable from those facts. Discussion transforms those facts into an argument for the broadest possible claim the data will legitimately support. Nevertheless, at this point a note of caution must enter. At both levels—the level of data generation and the level of data interpretation—

scientists must be constantly aware of the limitations of their methods and the fallibility of their inferences. Therefore, the broader the claim, the greater is the necessity for qualification. Scientists must constantly be aware that certainty is never the property of an individual. It is always the property of the community to which the individual belongs. Only this community can confer certainty—and the bestowal is always and necessarily provisional.

The stakes here are high for authors—whether or not their hard work over an extended period will receive any recognition from their peers. If they are successful, readers will accept their claim, tentatively, as knowledge and, in the best-case scenario, will pass it along to other researchers, turning it into an accepted claim and recognizing them as its owner, preferably by means of multiple citations in the literature. In the worst case, readers will reject the claim and ultimately discredit or question it at scientific symposia and informal gatherings or in the journal literature.

EXERCISES

Exercise 1

Disaster struck on January 28, 1986, when the space shuttle *Challenger* exploded a little over one minute after launch—with family, friends, and other spectators watching in horror. Subsequent investigation attributed the accident's cause to the large rubber O-rings (38 feet in diameter yet only 0.25 inch thick) in the booster rockets, rings that leaked because of unusual cold: the night before launch, the temperature had dropped to below freezing. The ambient temperature at launch was only 36° Fahrenheit.

Many articles and books about this tragic engineering failure have since appeared. One of the better known and more provocative is the centerpiece of Edward Tufte's *Visual Explanations* (1997). In it he assembled a table of key data culled from the five-volume report on the accident investigation issued by the Committee on Science and Technology of the House of Representatives. We reproduce a modified version here (table 2), showing the rocket casing analyses from the twenty-four previous launches of the space shuttle, all successful but not without problems. We order the table vertically by flight number and date of launch. The last two columns present the key data: O-ring temperature at time of launch and a damage index determined after the mission.

If you were preparing this table, would you have arranged it in the same way we have, with date of launch controlling the order? If not, why not? Is there any trend apparent from these data? How could you arrange the table to clarify the trend?

Next, create a scatter plot of the data with the temperature on the x-axis

TABLE 2. O-Ring temperature and damage index by space-shuttle flight number. Adapted from Tufte 1997.

Flight Number	Date	O-ring Temp. (°F)	Damage Index
1	4/12/81	66	0
2	11/12/81	70	*4*
3	3/22/82	69	0
4	6/27/82	80	?
5	4/12/82	68	0
6	4/4/83	67	0
7	6/18/83	72	0
8	8/30/83	73	0
9	11/28/83	70	0
41-B	2/3/84	57	*4*
41-C	4/6/84	63	*2*
41-D	8/30/84	70	*4*
41-G	10/05/84	78	0
51-A	11/8/84	67	0
51-C	1/24/85	53	*11*
51-D	4/12/85	67	*0*
51-B	4/29/85	75	0
51-G	6/17/85	70	0
51-F	7/29/85	81	0
51-I	8/27/85	76	0
51-J	10/03/85	79	0
61-A	10/30/85	75	*4*
61-B	11/26/85	76	0
61-C	1/12/86	58	*4*
51-L	*1/28/86*	*26–29*[a]	*(catastrophic failure)*

a. Forecasted temperature on day before launch of the *Challenger.*

and the damage index on the *y*. What does this transformation accomplish? Is the evidence from this scatter plot, which the skeptical managers never saw, strong enough to have convinced them to halt the countdown?

Read Tufte's analysis (Tufte 1997, 38–53) and compare it with yours. Do you agree with Tufte's harsh judgment of the engineers and managers involved? For a counterargument to Tufte's analysis and a defense of the engineers (but not the managers who ignored their reservations), you might also read "Representations and Misrepresentations: Tufte and the Morton Thiokol Engineers on the *Challenger*," by Wade Robinson et al. (2002).

Exercise 2

Here is a real-world medical puzzler meant to illustrate the difficulty of making sound comparisons within a research study or even between two studies published in respected journals. Two research groups tackled the same medical problem with opposite conclusions. Read both papers: Henschke et al. 2006 and Bach et al. 2007.

The problem they both studied was as follows: Is there any benefit to routine screening by computerized tomography (CT) for individuals at risk for lung cancer? Research Group A (Henschke et al. 2006) concluded yes, Research Group B (Bach et al. 2007), no.

Research Group A screened about 30,000 such people and identified 484 with lung cancer. The lung cancer patients then underwent the usual medical treatments depending on the severity of the disease. Research Group A then compared the ten-year survival of such patients with that of current lung cancer patients (no CT screening). The benefit appeared to be dramatic.

Research Group B screened about 3,000 such people and found 144 with lung cancer. These patients also underwent the usual treatments. The researchers compared the mortality rate (death due to lung cancer) for the screened patients with that predicted for lung cancer patients with no screening. They found no benefit from the screening.

Can you explain why the two groups reached different conclusions? Which study do you find more persuasive? In part, the answer is that earlier detection of a potentially lethal disease does not necessarily mean that the probability of the final deadly outcome will diminish. For one perspective on this puzzler, see a *New York Times* essay by H. Gilbert Welch, Steven Woloshin, and Lisa M. Schwartz, "How Two Studies on Cancer Screening Led to Two Results" (March 13, 2007).

CHECKLIST

Critical readers will be persuaded of the value of your research on the basis of three fundamental factors: Have you established an original problem worth solving, and what is your or your institution's track record for solving such problems? Have you developed a plausible strategy for solving the problem? Have you made the case that the results generated from having executed the method actually solved the problem in whole or part (in theory, you could publish an article to establish that some method does *not* solve a problem that many others are working on, but such articles appear infrequently)? Here is our checklist to help best make your case in results and their discussion:

- For results, when you arrange your tables and visuals in the order you have chosen, do they form the data backbone of the story about them that you want to tell, a backbone with no missing vertebra?
- For results, does your text make clear the limitations of your methods—limitations that might affect its factual status?
- For results, is your discussion limited to the immediate inferences you can make from your data?
- For any tables and figures, have you explained the meaning behind the displayed data, if it is not immediately obvious visually?
- For discussion, are the data from the results section integrated into an argument that makes the broadest claims that are legitimate, given the existing evidence? Could the significance of your data be made clearer or your argument stronger by a comparison (experimental vs. control, present vs. past, normal state vs. altered, experimental vs. theoretical, etc.)?
- For discussion, are these broad claims so qualified that they exemplify your best judgment that their strength is fine-tuned for likely acceptance by the community of scientists who are your audience? Have you made every effort to take into account possible criticisms of your fundamental argument by that discourse community?

5 Drawing Your Conclusions

In my beginning is my end . . .
In my end is my beginning.

T. S. Eliot, *Four Quartets* (1942)

The young Isaac Newton began his first scientific paper by announcing what he did and when he did it to the editor of *Philosophical Transactions* and its readers, members of the fledgling Royal Society of London: "To perform my late promise to you, I shall without further ceremony acquaint you, that in the beginning of the Year 1666 . . . I procured me a Triangular glass-Prisme, to try therewith the celebrated *Phænomena* of *Colours*" (Newton 1672). After having discharged his introductory promise in the subsequent pages, Newton ends his paper with a challenge to his audience:

> This, I conceive, is enough for an Introduction to Experiments of this kind; which if any of the *R. Society* shall be so curious as to prosecute, I should be very glad to be informed with what success: That, if any thing seem to be defective, or to thwart this relation, I may have an opportunity of giving further direction about it, or of acknowledging my errors, if I have committed any.

That paragraph makes for a graceful ending with which Newton apparently hopes to elicit critical feedback. When several readers mistook politeness for openness to criticism and actually did question his results and conclusions, Newton issued rebuttals in writing, oscillating from the standard neutral voice of science to the barely civil and even surly. At times, he seemed truly mystified that anyone would dare question a research project to which he had devoted so much time. Whether or not Newton later felt any qualms about publishing that closing statement, its challenge has remained implicit in all experimental scientific papers published since. Indeed, for a modern scientist to end a paper with such a statement today would be to state the obvious.

Typical Structure of a Scientific Conclusion

Our concern in this chapter is what typically appears in the modern scientific conclusion section. In parallel to Swales's three-step introduction, we identify three critical elements:

1. original claims supported by the evidence in the previous text
2. wider significance of those claims to the research territory under scrutiny
3. possible future work to validate or make use of the original claims

The first element reiterates and expands upon the authors' chief claims regarding the problem set in their introduction. The second places these claims within the broader context of current disciplinary knowledge and debate. The third makes a case for the continuing value of the authors' research program: few articles are ever the final word on a research problem. The conclusion also marks the authors' last chance to gain the approval of their readers, or at least drive home the main message. Each of these elements has a close relative in the introduction:

1. The first element concerns the solution to the problem. In the typical introduction we are given a brief glimpse of that solution; the conclusion usually presents a full statement of it.
2. The second element concerns the research territory. While introductions must show an awareness of the context of a specific research front, conclusions must be sensitive to the alterations in that context that the new claims to knowledge make.
3. The third element concerns the research solution's implications. While the introduction formulates a new problem about to be solved, the conclusion suggests future challenges generated from that solution.

As an example, we turn to the conclusion of an article on the evolution of hands and forelimbs from fish fins, based on the discovery of a transitional fossil called *Tiktaalik roseae* (Shubin, Daeschler, and Jenkins 2006). We first compare claims:

CONCLUSION	INTRODUCTION
The claim "The pectoral fins of the *Tiktaalik* reveal that development of robusticity and mobility of the distal skeleton was	*The claim anticipated* "The discovery of *Tiktaalik roseae* brings new data to bear on these issues [of fin-limb transition]. The material is re-

underway before the origin of tetrapods. . . . The distal endoskeleton of the *Tiktaalik* invites direct comparisons to the wrists and digits of limbed vertebrates."	markable for its phylogenetic position, three dimensional preservation and abundance."

We next compare treatments of the research territory:

CONCLUSION	INTRODUCTION
Impact on research territory "The pectoral skeleton of *Tiktaalik* is transitional between fish fin and tetrapod limb. Comparison of the fin with those of related fish reveals that the manus [human hand or quadruped forefoot] is not a *de novo* novelty of tetrapods; rather, it was assembled in fishes over evolutionary time to meet the diverse challenges of life in the margins of Devonian aquatic ecosystems."	*Impact on research territory anticipated* "A landmark event in vertebrate history is the transformation of fish fins into tetrapod limbs. Insights into this transition illuminate the biological mechanisms that generate major shifts in developmental genetics, skeletal structure and biomechanics."

Finally, we turn to the issue of future work:

CONCLUSION	INTRODUCTION
Research generated by solution of the original research problem " . . . Studies of the genetic basis of cartilage patterning in more basal actinopterygians or sarcopterygians [names for two types of prehistoric fish] may ultimately be more informative of the transformations that occurred in the Devonian period."	*Original research problem* "An impediment to understanding the fin-limb transition has been the nature of available evidence from the sister group of tetrapods. The closest living relatives of tetrapods—lungfish and coelacanths—either lack homologous elements to distal limb bones or are so specialized that comparisons with tetrapods are uncertain . . . "

In the typical conclusion, the three elements are not nearly so firmly entrenched as they are in the typical introduction. In fact, our survey of conclusions in twentieth-century scientific articles found less than 15 percent with all three elements and nearly 40 percent with none at all. Still, over

60 percent of articles did have a distinct conclusion section. Our rule of thumb is that the longer the article, the greater the need for some kind of discrete conclusion. But a great deal of variation is tolerated.

In the article on the fin-limb evolutionary transition, for example, although all the elements are present, they are not presented in the order we have designated as canonical. Instead, the original claim is summarized, followed directly by the statement of suggested lines of new research, followed in last place by a statement of wider significance. The authors chose to structure their conclusion that way because they wanted to close with a statement linking their new claim to a coherent, ongoing program of research with major implications for evolutionary theory. Even more drastic variations in conclusions are not uncommon. For example, the conclusion of an article on embryonic stem (ES) cells focuses on the potential uses of the authors' discovery to the exclusion of the other concluding elements (Thomson et al. 1998):

> Human ES cells should offer insights into developmental events that cannot be studied directly in the intact human embryo but that have important consequences in clinical areas, including birth defects, infertility, and pregnancy loss. . . . Screens based on the in vitro differentiation of human ES cells to specific lineages could identify gene targets for new drugs, genes that could be used for tissue regeneration therapies, and teratogenic [tumor-causing] or toxic compounds.

The vital importance of these applications accounts for this variation in form.

Claims as Theme and Variations

Statements dealing with original claims can appear in five locations in a scientific article: the title, abstract, introduction, results and discussion, and, last but not least, conclusion. That might appear to be overkill. But given the difficulties imposed by the complexities of modern scientific prose, we contend that most readers happily welcome continued reinforcement, as in this article in which the authors report on a tic-tac-toe playing computer built out of DNA (Stojanovic and Stefanovic 2003):

> *Full title*
> A deoxyribozyme-based molecular automaton
>
> *From abstract*
> We describe a molecular automaton, called MAYA, which encodes a version of the game of tic-tac-toe and interactively competes against a human oppo-

nent. The automaton is a Boolean network of deoxyribozymes that incorporates 23 molecular-scale logic gates and one constitutively active dexyribozyme arrayed in nine wells (3x3) corresponding to the game board.

From introduction

Here we report three-input deoxyribozyme-based logic gates i_AANDi_BANDNOTi_C and use them in a bottom-up approach as building blocks to construct a solution-phase Boolean molecular network, called MAYA, which plays a dynamic game of tic-tac-toe against a human opponent and never loses.

From results and discussion

We proceeded to construct three-input deoxyribozyme-based i_AANDi_BANDNOTi_C gates . . . with eight possible states, one of which is active. . . . We describe below how the logic representation of a tic-tac-toe strategy was transformed so that such three-input gates sufficed. . . . The game tree has 19 possible games, and we have played all 19 against MAYA at least four times each. In every case, MAYA performed to specification and was never defeated—as expected, because it implements a perfect strategy.

From conclusion

The deoxyribozyme-based automaton MAYA precisely executes a linear program encoded through a spatial distribution of nucleic acid catalysts behaving as logic gates. In a total of > 100 games played, we detected no erroneous moves by MAYA; in other words, under all tested conditions the fully activated gate was always more fluorogenically active than an assembly of several partially active gates. . . . No assembly of biomolecules that can autonomously play a dynamic game has been described before this report.

The title of this article captures the main claim in an abbreviated noun phrase. The abstract expands that claim in two complete sentences. The solution statement from the introduction rephrases the abstract. The discussion of results makes assertions analogous to those in the introduction in the context of a detailed description of the building and testing of the automaton. Finally, the conclusion highlights the main claims in a new context. While some repetition occurs from passage to passage, each gives us some unique information about the solution in a different context. By simply scanning the article for such claim-staking statements, a reader will get a good picture of what the authors accomplished. And many readers perform that sort of selective reading and stop at that point.

For that reason, we recommend that once you have written a first draft of an article, you scroll through it looking for statements to check whether your main message comes though clearly. Are they simply repetitive, or does each one add something new to the message?

Regardless of whether you have an actual discrete conclusion section, your article must be so organized that its main point or claim remains firmly rooted in your readers' consciousness. This is never a matter of mere repetition. It is a theme with variations. In Frederic Rzewski's piano composition "The People United Will Never Be Defeated!" the theme from a Chilean revolutionary anthem is stated, followed by thirty-six variations, followed finally by a repetition of the original theme. The repetitions of the claim in a scientific article are analogous.

Casting the Net of Wider Significance

In chapter 1 we recommended that you ask the "so what?" question when introducing your research problem. Now we suggest that you ask this same question at the very end of your article. After having presented your solution to the problem, put yourself into the shoes of a highly critical reader and ask, "So what? Why should I care about this solution?" You can answer that question in either the introduction or conclusion.

In the final paragraph of a 1927 paper on quantum mechanics, for instance, Werner Heisenberg addresses the provocative philosophical implications of his famous equation concerning the uncertainty relation between the position and momentum of a subatomic particle:

> If one assumes that the interpretation of quantum mechanics is already correct in its essential points, it may be permissible to outline briefly its consequences of principle. . . . [E]verything observed is a selection from a plenitude of possibilities and a limitation on what is possible in the future. As the statistical character of quantum theory is so closely linked to the inexactness of all perceptions, one might be led to the presumption that behind the perceived statistical world there still hides a "real" world in which causality holds. But such speculations seem to us, to say it explicitly, fruitless and senseless. Physics ought to describe only the correlation of observations. One can express the true state of affairs better in this way: Because all experiments are subject to the laws of quantum mechanics, and therefore to [the equation I have just given], it follows that quantum mechanics establishes the final failure of causality.

To return to the paper on a DNA computer invincible in tic-tac-toe, any inquisitive reader would naturally want to know: What's the point? Why all the bother? Milan Stanjanovic and Darko Stefanovic oblige with an answer in their last two sentences. They postulate that a more complex version of their computer could one day "be used in synthetic biology to learn more about genetic, regulatory and metabolic systems networks that culminate in complex decisions such as division, differentiation and movement. Another application . . . might be *in vivo* [in living matter] computation networks and cells with engineered properties [for fighting diseases]."

Announcing Future Work

Implicit at the end of any play is the unspoken fate of the remaining characters. In a tragedy, those who remain alive must try to recover from some disaster; in a comedy, they have overcome adversity and, we presume, will live happily ever after. Only on rare occasions do authors append a coda spelling out what will happen to their characters in an imagined future. But unlike literary works, scientific articles often specify what the future might hold in store.

Almost all scientific papers have happy endings, with the authors having solved some problem to their satisfaction and, they hope, to the satisfaction of their readers. In general, solving a significant research problem will create new problems that the same authors or others can tackle. Researchers around the globe are still working on problems that can be traced to Charles Darwin's *On the Origin of Species.* The same can be said for Albert Einstein's four articles published in 1905, his miraculous year, or Watson and Crick's solution to the structure of DNA. A similar statement could be made about human stem-cell research during the 1990s. But not all scientific articles have or need a statement on exactly what future work might involve. How do you decide whether your particular paper would benefit from one?

First, ponder what you might say about the next step forward. Then ask yourself a few questions. Would your intended readers be wondering what's the next step? Would such a statement help them better appreciate the significance of your having solved the original problem? Even if your answer to both questions is yes, you may still want to exercise caution. Do you want to risk handing your competitors information that they might exploit at your expense? Having answered these questions, you will be better positioned to decide whether you want to conclude with a statement concerning future research. Here is an example of such a statement from a classic article on child psychology (Bandura, Ross, and Ross 1961), one that nicely avoids the risk of preemption from competitors:

> The experiment reported in this paper focused on immediate imitation [by the child] in the presence of the [adult] model. A more crucial test of the transmission of behavior through the process of social imitation involves

the generalization of imitative responses to new situations in which the model is absent. A study of this type, involving the delayed imitation of both male and female aggressive models, is currently under way.

We would strongly discourage a merely perfunctory statement about future work. Readers of the scientific literature take a dim view of statements of the obvious, such as "Work will continue on this research project if funding is available."

The Very End

The final chapter to Darwin's *On the Origin of Species* opens with this sentence: "As this whole volume is one long argument, it may be convenient to the reader to have the leading facts and inferences briefly recapitulated." The chapter that follows includes the standard three elements of a conclusion section in canonical order. Darwin first recapitulates his main claims, covering both the difficulties posed by his theory and the overwhelming evidence in support of it. He then establishes his theory's place within natural history, the disciplinary ground out of which evolutionary biology grew. Finally, he speculates on the "grand and almost untrodden field of inquiry" that his research will open, a future he foresees reaching into diverse fields such geology and psychology. In the very last paragraph of the last chapter, Darwin brings his work to a rousing close:

> It is interesting to contemplate an entangled bank, clothed with many plants of many kinds, with birds singing on the bushes, with various insects flitting about, and with worms crawling through the damp earth, and to reflect that these elaborately constructed forms, so different from each other, and dependent on each other in so complex a manner, have all been produced by laws acting around us. These laws, taken in the largest sense, being Growth with Reproduction; inheritance which is almost implied by reproduction; Variability from the indirect and direct action of the external conditions of life, and from use and disuse; a Ratio of Increase so high as to lead to a Struggle for Life, and as a consequence to Natural Selection, entailing Divergence of Character and the Extinction of less-improved forms. Thus, from the war of nature, from famine and death, the most exalted object which we are capable of conceiving, namely, the production of the higher animals, directly follows. There is grandeur in this view of life, with its several powers, having been originally breathed into a few forms or into one; and that, whilst this planet has gone cycling on according to the fixed law of gravity, from so simple a beginning endless forms most beautiful and most wonderful have been, and are being, evolved.

Darwin's final sentence works so powerfully by contrasting permanence and change: the earth's unchanging revolution around the sun compared with the "most beautiful and most wonderful" life forms evolving on earth over many hundreds of millions of years.

With rare exception, modern scientific articles, even those reporting major breakthroughs, do not end in Darwin's grand style. Still, unless they are very short, they benefit from a closing statement that recapitulates "leading facts and inferences" or looks forward to future research. Here are a few examples of closing sentences from significant articles published over the last century:

> The evidence presented supports the belief that a nucleic acid of the deoxyribose type is the fundamental unit of the transforming principle of Pneumococcus Type III. (Avery, MacLeod, and McCarty reporting that genes are made of DNA, 1944)
>
> In view of its success with equilibrium properties, it may be hoped that our theory will be able to account for these and for other so far unresolved problems. (Bardeen, Cooper, and Schrieffer reporting a new theory on how superconductivity works in metals at temperatures near absolute zero, 1957)
>
> The results also suggest that superconductivity at temperatures greatly exceeding 40 K is achievable in LBCO and related systems through the fine tuning of the sample parameters by physical and chemical means. (Chu et al. reporting on a new superconducting material [lanthanum-barium-copper oxide, or LBCO] at "high" temperatures of 40 K, 1987)

Good first impressions pique readers' interest. Good last impressions stick in their minds of readers long after they have set the scientific journal aside or logged off the World Wide Web.

EXERCISE

Go to the 1953 *Nature* article by Watson and Crick on the structure of DNA (www.nature.com/nature/dna50/archive.html). It is very short (fewer than 1,000 words) and has no distinct conclusion section. See if you can write one with all three components in the canonical order.

Our Answer

Solution

We have developed a credible model for the structure of DNA. It consists of two helical chains coiled around the same axis. The chains are held together by the purine and pyrimidine bases in accord with Chargaff's rule.

Wider significance
The double helical structure has a simple copying mechanism for the manufacture of genetic material.

Future research
The structure is roughly compatible with the experimental data, but the details still need to be checked against more exact results.

CHECKLIST

Once you have a complete first draft for a new research article, we suggest that you ask yourself the following questions:

- Did you rephrase your main claim at strategic points throughout the article?
- Did the rephrasing of the main claim later in the article, and especially in the conclusion, differ according to context from statements earlier in the paper?
- In formulating your conclusion, have you asked yourself why readers should consider solving your problem worthwhile?
- In your conclusion, do you look forward to the research your solution will generate? or to possible practical applications?
- Does your conclusion have all three elements in the canonical order? If not, is there a good reason that an element is missing or out of order?

6 Framing Your Methods

The modern methods section is as varied as the procedures, materials, and theoretical principles employed within the numerous specialties that populate science. Yet however diverse the content, all methods sections share the same fundamental purpose: to inform the reader by what means the authors solved the problem stated in the introduction.

Of all the sections in a scientific article, the methods section is probably the least read, for reasons we will divulge momentarily. Yet its inclusion is essential to the overall argument. As sociologist of science Harry Collins notes in his monumental *Gravity's Shadow: The Search for Gravitational Waves* (2004), "The discussion of methodology provides a warrant for a study's findings"; it forms the basis for why scientist-readers ought to believe that the findings solve the stated problem. For that reason, it often appears immediately after the introduction; for other reasons, it sometimes appears after the conclusion or is woven into the figure and table captions. Journal style dictates where the methods section appears, though its usual place is the logical one—after the introduction and before the results.

We find it helpful to think of methods sections in experimental articles as roughly divided along chronological lines:

- *Preparation for carrying out experiments to solve the research problem mentioned in the introduction*
  This step often includes descriptions of any special treatments to the objects of the experiment, as well as their principal characteristics such as dimensions, weight, volume, and composition. It also typically includes some description of the equipment and other relevant information about the site of the experiment.
- *Actual experiments*
  This step includes actions applied to the objects at the experimental site, along with conditions (temperature, duration, pressure) that might alter the results if they changed and the experiment were to be repeated.

- *Analysis of information produced by the experiment*
  This step includes procedures applied to the objects to generate data for analysis either during or after the actual experiment.

This order is *not* meant to reflect a prototypical structure, as is the case with the introduction and the scientific article as a whole. Methods sections can almost never be so neatly divided. Moreover, so simple a narrative structure seldom meshes with the complexity of a series of experiments that stretches over weeks or years. Still, we find it useful as a general guide for thinking about the methods section. You should also be aware that certain types of article—those making observations about nature as opposed to measurements or calculations and purely theoretical articles, for example—typically do not have a distinct methods section. Reviews of the literature, of course, always lack one.

Later in this chapter, we will be discussing these three narrative elements in greater detail. But we first address an important question.

How Detailed Should the Methods Section Be?

The methods section of the scientific article is organized somewhat like the recipe in a cookbook. This is how Julia Child (2000) instructed her readership on the preparation of leeks for leek and potato soup:

> Trim off the root ends, keeping the leaves attached. Cut off tops so that the leeks are 6 to 7 inches long. Starting ½ inch from the root and keeping leaves attached, slit each leek lengthwise in half and then in quarters. Wash under cold running water, spreading the leaves apart to rinse off all dirt. Leeks can be braised whole or sliced crosswise into pieces for soup. To julienne, cut leeks crosswise into 2-inch pieces, press leaves flat, and slice lengthwise into matchsticks.

Compare that passage with biochemists E. G. Bligh and W. J. Dyer's (1959) description of an important new method for extracting and purifying lipids from animals:

> The following procedure applies to tissues like cod muscle that contain 80 ± 1% water and about 1% lipid. Each 100-g sample of the fresh or frozen tissue is homogenized in a Waring Blender for 2 minutes with a mixture of 100 ml chloroform and 200 ml methanol. To the mixture is then added 100 ml chloroform and after blending for 30 seconds, 100 ml distilled water is added and blending continued for another 30 seconds. The homogenate is filtered through Whatman No. 1 filter paper on a Coors No. 3 Büchner

> funnel with slight suction. Filtration is normally quite rapid and when the residue becomes dry, pressure is applied with the bottom of a beaker to ensure maximum recovery of solvent. The filtrate is transferred to a 500-ml graduated cylinder, and, after allowing a few minutes for complete separation and clarification, the volume of the chloroform layer (at least 150 ml) is recorded and the alcoholic layer removed by aspiration. A small volume of the chloroform is also removed to ensure complete removal of the top layer. The chloroform layer contains the purified lipid.

Both passages proceed in step-by-step fashion to explain how to get from some set of ingredients to a finished product—edible in the case of Julia Child, useful for biological research in the case of Bligh and Dyer. Most instructional manuals—whether concerned with the setup of a new personal computer or with the building of a model airplane—conform to this basic narrative style. Yet despite similarities, there is an important difference between these two sorts of narrative: the level of detail involved.

Conventional wisdom holds that the methods section should be so detailed that a researcher working in the same area can repeat the reported research with the same results. It depends. Readers expect such elaborate details when the authors are reporting on a new research method that others might copy or modify, as is the case for the article by Bligh and Dyer, or when they are making claims they know others will find important or controversial and might want to verify, as was notoriously the case when Stanley Pons and Martin Fleischmann claimed in 1989 to have fathered a "cold fusion" reaction in a tabletop device. But relatively few experiments get repeated just for the sake of repetition, because replication in modern science is often a complex, time-consuming, and expensive business, with little reward whether the experimenter succeeds or fails.

In our view, the more important purpose of the methods section is to make the case before experts that executing the methods stated therein constitutes a plausible strategy for solving the research problem presented in the introduction. The remainder of this chapter concerns how we think you can best accomplish that end.

Advance Notice

Many readers of scientific articles do not much care about the nitty-gritty details of the methods and will not scrutinize that section very carefully, if at all. They normally take for granted that the authors knew what they were doing and did what they have stated in print. They take as their guarantee the fact that the methods passed muster with the journal referees and editors during the prepublication review. Censorious readers, on the other hand, will immediately question the validity of the results when the methods section is sketchy. They want minute detail.

Whatever the case, all interested readers benefit from some general statement on the methods applied to solve the stated problem. What in a nutshell did the authors do in the lab or field or at their office computers to solve the problem? That statement can appear in the introduction or abstract or at the start of the methods section.

In the article on the imitation of aggressive behavior by children, Bandura and his collaborators (1963) begin their abstract with a problem statement followed by a synopsis of their experimental approach for solving it, printed here in italics:

> To test the hypothesis that exposure of children to film-mediated aggressive models would increase the probability of Ss' [the subjects'] aggression to subsequent frustration, *1 group of experimental Ss observed real-life aggressive models, a 2nd observed these same models portraying aggression on film, while a 3rd group viewed a film depicting an aggressive cartoon character. Following the exposure treatment, Ss were mildly frustrated and tested for the amount of imitative and nonimitative aggression in a different setting.*

One of the major failings in many scientific articles is that the reader is knee deep in a bewildering array of materials and methods without ever being given some overview of how they connect to the introductory problem. Armed with advance knowledge of how the authors went about solving their research problem, however, interested readers are much better prepared for understanding the rest of the article.

Narrative Structure

As we just mentioned, good methods sections tell a story with beginning, middle, and end, interrupted at times by a well-justified digression. For an experimental article, that typically means describing what was done to set up the experiment, to carry it out, and to gather and analyze the results. Here are those elements in the methods section of Bandura, Ross, and Ross in the psychology test lab (we have added headings for clarification):

> *Preparations*
> The subjects were 36 boys and 36 girls enrolled in the Stanford University Nursery School. They ranged in age from 37 to 69 months, with a mean age of 52 months.

Two adults, a male and female, served in the role of model, and one female experimenter conducted the study for all 72 children.

Subjects were divided into eight experimental groups of six subjects each and a control group consisting of 24 subjects. Half the experimental subjects were exposed to aggressive models that were subdued and nonaggressive in their behavior. These groups were further subdivided into male and female subjects. . . .

Execution of the experiment

In the first step in the procedure subjects were brought individually by the experimenter to the experimental room and the model, who was in the hallway outside the room, was invited by the experimenter to come and join the game. . . . [T]he experimenter escorted the model to the opposite corner of the room which contained a small table and chair, a tinker toy set, a mallet, and a 5-foot inflated Bobo doll. . . .

With subjects in the *nonaggressive condition,* the model assembled the tinker toys in a quiet subdued manner totally ignoring the Bobo doll.

In contrast, with subjects in the *aggressive condition,* the model began by assembling the tinker toys but after approximately a minute had elapsed, the model turned to the Bobo doll and spent the remainder of the period aggressing toward it. . . .

Data acquisition and analysis

The subjects were rated on four five-point rating scales by the experimenter and a nursery school teacher, both of whom were well acquainted with the children. These scales measured the extent to which subjects displayed physical aggression, verbal aggression, aggression toward inanimate objects, and aggressive inhibition. . . .

Fifty-one subjects were rated independently by both judges so as to permit an assessment of interrater agreement. The reliability of the composite aggression score, estimated by means of the Pearson product-moment correlation, was .89.

The composite score was obtained by summing the ratings on the four aggression scales.

We have purposely selected paragraphs and rearranged the order to conform to our simplified structure. In the published methods section, the third item followed the first. That deviation aside, the overall narrative structure of the complete methods section essentially conforms to the three steps of our canonical structure. You can infer that from the actual headings used (numbered items in italics below, arranged in columns under the typical three steps):

PREPARATIONS	EXPERIMENTS	ANALYSES
1. Subjects *2. Experimental Design*	*3. Experimental Conditions* *4. Aggression Conditions* *5. Test for Delayed Imitation*	*6. Response Measures*

The first two headings cover preparations; the next three, the actual experiments; and the last one, data acquisition and analysis. The results section that immediately follows the last heading is loosely organized around the order of the "response measures" (imitation of physical aggression, verbal aggression, nonaggressive verbal responses). Good methods sections often suggest an obvious organizing principle for the results and discussion that follow.

We find two common failings in the narratives of methods sections. Sometimes authors begin by listing the materials used, just as in a cookbook recipe, but fail to account for all of these materials when describing the execution of the experiment. At other times, the authors mention methods applied or equations employed but neglect to make clear their findings in results and discussion. In other words, just as you need to tie your methodological approach to your introductory problem, you need to tie your methods for acquiring and analyzing data to the presentation of results. By those measures, it appears to us that Bandura, Ross, and Ross succeeded admirably.

Rationale for Choices

Linguist John Swales (1990) has noted that the modern methods section typically has "little statement of rationale or discussion of the choices." In general that is true for articles about investigations in which the authors use previously published and accepted methods in their discipline. Indeed, the convention is to direct your reader to relevant sources by means of citation. However, we do not believe Swales's assertion is true for methods that extend the frontiers of sample preparation, measurement, and analysis. Good writers anticipate points where knowledgeable readers will question a choice of method or material or type of analysis; having done so, they add some justification. This tactic not only informs the reader but also defuses potential criticisms.

There is no better illustration of this principle than the famous article by Avery, MacLeod, and McCarty (1944) establishing that DNA is responsible for transmitting genetic traits. They demonstrate this principle in a pneumonia-inducing bacterium (pneumococcus type III) because it represented "the most striking example of inheritable and specific alterations in cell structure and function that can be experimentally induced and are reproducible." In the very first paragraph under the heading "Experimental" we learn that the authors will be crossing the border into previously unexplored methodological territory:

> Transformation of pneumococcal types *in vitro* requires that certain cultural conditions be fulfilled before it is possible to demonstrate the reaction even in the presence of a potent extract. Not only must the broth medium be optimal for growth but it must be supplemented by the addition of serum or serous fluid known to possess certain special properties. . . . Each constituent of this system presented problems which required clarification before it was possible to obtain consistent and reproducible results.

Reading on, we learn that

> [i]n the first successful experiments on the induction of transformation *in vitro,* Dawson and Sia . . . found it was essential to add serum to the medium. . . . In the present study human pleural or ascetic fluid has been used almost exclusively. It became apparent, however, that the effectiveness of different lots of serum varied and that the differences observed were not necessarily dependent upon the content of R antibodies, since many sera of high titer were found to be incapable of supporting transformation. This fact suggested that other than R antibodies are involved.

Unlike a typical cookbook recipe, the methods section can incorporate some explanation of the rationale behind the choices that were made. So in this passage, and throughout the long methods section in this article, we learn not only what the authors did but why they did it and what major difficulties they overcame.

A Matter of Style

The preferred verb voice in the modern methods section is the passive, that is, some form of the verb *to be* combined with a past participle verb, as in *was found.* We repeat the Bligh and Dyer (1959) paragraph to illustrate, with the passive verbs marked by italics:

> The following procedure applies to tissues like cod muscle that contain 80 ± 1% water and about 1% lipid. Each 100-g sample of the fresh or frozen tissue *is homogenized* in a Waring Blender for 2 minutes with a mixture of 100 ml chloroform and 200 ml methanol. To the mixture *is then added* 100 ml chloroform and after blending for 30 seconds, 100 ml distilled water *is added* and blending *[is] continued* for another 30 seconds. The homogenate *is filtered* through Whatman No. 1 filter paper on a Coors No. 3

> Büchner funnel with slight suction. Filtration is normally quite rapid and when the residue becomes dry, pressure *is applied* with the bottom of a beaker to ensure maximum recovery of solvent. The filtrate *is transferred* to a 500-ml graduated cylinder, and, after allowing a few minutes for complete separation and clarification, the volume of the chloroform layer (at least 150 ml) *is recorded* and the alcoholic layer *[is] removed* by aspiration. A small volume of the chloroform *is also removed* to ensure complete removal of the top layer. The chloroform layer contains the purified lipid.

The passive voice allows writers to avoid mentioning a human agent. An object, process, or concept appears in the subject position, avoiding the pronoun *we,* the typical agent of the actions in science. Here is the beginning of the Bligh and Dyer paragraph rewritten in that style: "*We applied* the following procedure to tissues like cod muscle that contain 80 ± 1% water and about 1% lipid. *We homogenized* each 100-g sample of the fresh or frozen tissue in a Waring Blender for 2 minutes with a mixture of 100 ml chloroform and 200 ml methanol." There is nothing grammatically wrong with *we* in the subject position in methods sections or anywhere else for that matter. However, for at least the last century science has favored the passive style used by Bligh and Dyer. This style has the advantage of keeping the objects of the experiment in the important subject position and gives more variety to the subject position.

Cookbooks like Julia Child's typically have mostly imperative verbs, as in "*trim off* the root ends," where the subject *you* is understood. Here is the beginning of the Bligh and Dyer passage converted to that style: "The following procedure applies to tissues like cod muscle that contain 80 ± 1% water and about 1% lipid. *Homogenize* each 100-g sample of the fresh or frozen tissue in a Waring Blender for 2 minutes with a mixture of 100 ml chloroform and 200 ml methanol. *Add* 100 ml chloroform to this mixture. *Blend* it for 30 seconds. *Add* 100 ml distilled water and continue blending for another 30 seconds." Some methods sections actually follow this style, and there is nothing wrong with doing so. But this style works best when one is giving step-by-step instructions for others to follow, as cookbooks typically do. Methods sections normally serve a different end as building blocks in an overall argument.

EXERCISE

Here is an exercise you can do to test or refine your skill in writing methods sections. Take a cooking recipe and convert it into the style of a typical scientific methods section (with verbs in the passive voice). You should feel free to be creative and embellish. We begin with an easy example involving the preparation and taste testing of rice.

Rice Recipe

Combine in a 2-quart saucepan 1 cup of rice and 2 cups of water. Stir lightly, bring to a rolling boil, and reduce to simmer (low boil). Cover with a tight-fitting lid and simmer for 15 minutes. Remove from heat and allow to stand for 5 minutes. Serves three.

Translation into Scientific Method

An experiment was set up in a laboratory with a Maytag oven (Model 4532). Two areas were reserved, one for material preparation, the other for testing. To begin, 0.24 liters of *Oryza sativa* and 0.47 liters of tap water were poured into an open metallic pot with volume of 2 liters. This mixture was heated under an open flame on the oven, stirred gently, and brought to a boil (100 °C at standard pressure); then the temperature was immediately reduced to yield a simmering liquid. The pot was then covered with a tight-fitting lid. This pressurized lower temperature was maintained for 15 minutes. This heat treatment saturates the *Oryza sativa* grains and evaporates the remaining water. To avoid excessive agglomeration of these grains, at least 5 minutes of cooling in the closed pot is needed.

Immediately thereafter, 0.2-liter portions of the resulting substance were sampled in the testing area by three individuals (one adult male, one adult female, and one female child) with a sterilized eating implement made of silver-coated steel. The analysts rated their portions for texture, smell, and taste on a scale of one to ten. The results for each measure were averaged, analyzed by standard statistical techniques (described in reference 1), and compared with similar tests for other substances reported earlier.

CHECKLIST

Here are some questions to ask yourself after you have written a research article with a methods section:

- Are all three methodical threads (preparations, experiment, data gathering and analysis) adequately covered?
- If you have a subsection listing materials to be used, do you explain in a subsequent section of your article how they are actually used?
- Is the information in your methods consistent with your results and discussion? For example, if you have specified an equation for calculating results are the results of that equation presented in an appropriate section?
- Is the order of information in your methods section consistent with that in your results and discussion? For example, if your methods section has separate theoretical and experimental parts in that order, then read-

ers expect that your results and discussion will reflect that expository decision.

- Are your methods standard practice or cutting edge? If the latter, the reader might benefit from some explanation of key problems encountered or of the rationale behind important choices made. Be advised, however, that readers do not appreciate being informed of minor problems commonly encountered in similar research or being given lengthy explanations for obvious choices.
- Will the reader benefit from a brief summary of your methods in a section other than that devoted to methods? Would this summary fit better in the abstract or introduction?
- Are the placement and level of detail of the methods section consistent with the style of the journal to which you intend to submit your article?
- Finally and most important, will readers find your methods a plausible strategy for solving your particular research problem?

7 Distributing Credit

Science is no exception to the rule that creators desire credit for their creations. One might counter that some early scientific journals favored giving no specific credit for authorship other than perhaps the author's initials. But that self-effacing practice did not last long.

Modern scientific articles reserve three places for distributing credit: list of authors, list of references, and acknowledgments. The first typically appears up front after the title, the other two at the very end. In this chapter, we cover why these items are important and offer some advice on their construction.

Establishing Ownership

The byline to a scientific article establishes two sorts of responsibility: personal and institutional. Here is a typical byline appearing at the head of an important 2006 article in *Nature* ("The Pectoral Fin of *Tiktaalik roseae* and the Origin of the Tetrapod Limb"):

Neil H. Shubin,[1] Edward B. Daeschler[2] and Farish A. Jenkins, Jr.[3]

You will find the meaning of the superscripts in the article's endnotes:

1. Department of Organismal Biology and Anatomy, The University of Chicago, Chicago, Illinois 60637, USA
2. Academy of Natural Sciences of Philadelphia, 19th and Benjamin Franklin Parkway, Philadelphia, Pennsylvania 19103, USA
3. Department of Organismal and Evolutionary Biology and Museum of Comparative Zoology, Harvard University, Cambridge, Massachusetts 02138, USA
   Correspondence to: Neil H. Shubin

Some journals place the authors' institutional affiliations at the article head; others place them in footnotes still others in endnotes, as above. Whatever the style, the implication is the same. The authors as well as their research institutes will bask in the glow of a big success; they will also wilt in the shadows cast by spectacular failures or skullduggery.

The list of authors establishes the ownership of a particular intellectual property. Those authors will be named in citations by others who use that intellectual property in their own research. Ironically, this list does not tell us who actually wrote the paper. Almost all papers these days have multiple authors, on rare occasion numbering in the hundreds. It is not uncommon for one of the authors to draft the whole document; thus some "authors" may have contributed little if anything to an article's composition. At other times, different authors write different sections. And in some instances, a person not even listed as an author prepares the first draft.

Given that the author list tells us little about who actually wrote a given multi-authored scientific paper, what does scientific authorship mean? Let's put that question more in terms of someone faced with composing a list of authors: what criteria should he or she use in deciding who belongs on the list and in what order they should appear?

In our view, "authorship" implies that the named individuals played a major role in arriving at the new knowledge being presented and are willing to be held accountable should the article come under critical attack or, worse, prove false or fraudulent. Ideally that would involve participation in all phases of the research process: defining the research problem, assembling the resources to solve the problem, defining and carrying out the method to solve it, analyzing the results and formulating a solution, and, finally, composing a paper to communicate the research results to others. In reality, the complexity and compartmentalization of work arrangements in the modern scientific world dictate that contributing to a few of those tasks generally suffices for authorship.

This broad scope has definitely diluted the achievement of authorship. What still remains prestigious is recognition as principal or lead investigator. Nobel laureate Christiane Nüsslein-Volhard reports that she agreed to be listed second on the article resulting from her doctoral thesis. Her supervisor at the time felt that a male colleague who had started the research project and had a family to support should appear first to boost his career. She did not object at the time but thinks back ruefully on the incident: "I could still foam: I get so angry about it" (quoted in Dreifus 2006).

Author lists normally signify the principal investigator by order of appearance, though practices differ among research disciplines, journals, and even research teams. The most common style is to place the principal investigator first, and if the research supervisor differs from the principal investigator, to insert the research supervisor's name either last or second. Other lists order author names alphabetically but usually include an asterisk pointing to a footnote naming the person to whom queries should be addressed; that person is most often the principal author or supervisor. Some research groups avoid conflicts about who should appear first on author lists by rotating the honor among their members.

Owing to the importance of credit to career advancement, controversies arise concerning authorship. A case in point is a controversy surrounding the first-ever cloning of a mammal from an adult cell, a sheep named Dolly. The first author of five, Ian Wilmut, acted as the supervisor of the research team but did not participate in the actual cloning experiments and did not draft the original paper. Yet because he was first "author" and supervisor, he received much of the resulting worldwide acclaim. When this situation became public, some in the scientific community voiced complaints about "honorary authorship." Others countered, with some justification, that the coach of a winning sports team deserves credit along with the players (Stafford 2006). Whatever the merits of the analogy, most would question the listing of the author who did the most work, Keith Campbell, last. Still, listing authors by degree of participation is only an informal practice that evolved haphazardly over time, not a requirement.

Perhaps it should be a requirement. To avoid the Wilmut problem, some journals, such as the *New England Journal of Medicine,* ask authors to spell out in the acknowledgments exactly who did what. That policy may seem an ideal solution to the problem of establishing ownership of the different stages of research. But even there, issues can arise as to who deserves credit for which parts of the research.

We have three suggestions for devising an author list:

- Settle questions regarding whom to include and in what order as early as possible in the research process.
- Consider specifying in a footnote or acknowledgments the role each author played.
- Familiarize yourself with the customs surrounding authorship in your discipline and even the journal to which you plan to submit a paper.

(Note: In the interest of full disclosure, both authors contributed equally to the conception and writing of the present book, drawing in the case of Harmon on more than a quarter-century of practice and in the case of Gross on nearly a half-century of scholarship and teaching.)

Acting Ethically

Readers have certain expectations for the authors of a research article or report of any kind. First and foremost, they expect that authors have done exactly what they say they did. They assume that no data that might have cast doubt on their conclusions have been "lost" or "forgotten." And, it should go without saying, they assume that no data were fabricated out of thin air or deliberately fudged.

Second, readers expect that none of the authors has a conflict of interest that unduly favored one outcome over another. In a sense, all professional researchers have a vested interest in a certain outcome, if only to win the attention of peers and keep their research projects funded from year to year. All knowledgeable readers understand that point. But readers, especially referees of submitted manuscripts, do need to know about other possible conflicts, for example, whether authors testing the efficacy of a drug developed by a pharmaceutical company have any past or present financial entanglements with that company. For that reason, many research journals, institutions, and professional organizations now require authors to disclose any possible conflicts of interest.

Third, readers expect that the authors are not directly or indirectly taking credit for someone else's intellectual property—whether it be a knowledge claim, a new research method, a figure or table of data, or a passage of scientific prose. Readers expect citations to be affixed to any important technical information derived from an outside source and quotation marks to enclose any sentences copied from others.

Fourth, readers expect that the authors, if questioned about their findings, would be able to retrieve the relevant computer files and laboratory notebooks that support any claims they are making. Funding agencies do occasionally (sometimes frequently) audit research they sponsor, and research that comes under suspicion for whatever reason or circumstance can even be subjected to intense legal scrutiny.

We recommend that responsible authors, before hitting the send button for any e-mail conveying a new manuscript intended for publication, consider whether they might have failed on any of those counts. A mistake could damage a career permanently and cast a permanent shadow over the research institute listed after the author names.

Giving Credit by Citation

Scientists increase their reputations not only by authorship in prestigious journals but also by means of the accumulation of citations in articles by their fellow scientists. In the theory of scholar and public intellectual Georg Franck (2002), citation is the coin by which means scientists pay other scientists for using their intellectual property. The more citations scientists receive from others, the greater their accumulated wealth—that is, the greater their reputation. In competition with others, scientists vie for the scientific community's attention. This system encourages scientists to pursue research that might be used by other members of the scientific community in a productive way. So the modern reference list has importance far beyond the presentation of dry bibliographic information. And authors have a responsibility to pay their intellectual debts for using someone else's property—even for the purpose of disowning it. They also have an obligation to read all the important sources on their topic, whether retrievable on the Web or not. Prudence would dictate this, if ethics did not.

Our own research has shown that the modern scientific article typically has twenty to thirty references, though the total can run into the hundreds in very long articles. Those references typically serve three main purposes.

- They cite earlier work needed to establish the intellectual context out of which the research problem arose.
- They list the sources of details on methods so that the authors can merely summarize what they did.
- They cite earlier work that the authors wish to refute or use as supporting evidence for their findings.

In general, authors focus on citing recent pertinent work and assume considerable prior knowledge of the subject matter. The great majority of references are no more than ten to fifteen years old.

We reproduce below the reference list from a short article reporting a newly discovered large molecule of pure carbon, initially named "buckminsterfullerene," later shortened to "fullerene" (Kroto et al. 1985):

1. Heath, J. R. *et al., Astrophys. J.* (submitted).
2. Dietz, T. G., Duncan, M. A., Powers, D. E., & Smalley, R. E. *J. chem. Phys.* 74, 6511–6512 (1981).
3. Powers, D. E. *et al., J. phys. Chem.* 86, 2556–2560 (1982).
4. Hopkins, J. B., Langridge-Smith, P. R. R., Morse, M. D., & Smalley, R. E. *J. chem. Phys.* 78, 1627–1637 (1983).
5. O. Brien, S. C. *et al., J. chem. Phys.* (submitted).
6. Rohfing, E. A., Cox, D. M., & Kaldor, A. J. *J. chem. Phys.* 81, 3322–3330 (1984).
7. Marks, R. W. *The Dymaxion World of Buckminster Fuller* (Reinhold, New York, 1960).
8. Heath, J. R. *et al., J. Am. Chem. Soc.* (in the Press).
9. Herbig, E., *Astrophys. J.* 196, 129–160 (1975).

This reference list cites nine documents for an article only two pages long. References 1 and 6 refer to similar earlier work out of which the present study emerged. References 2 to 5 give details on experimental method. Reference 7 refers to a book on Buckminster Fuller, whose geodesic dome bears a striking resemblance to the structure of the sixty-atom carbon cluster the present authors discovered (hence its name). References 8 and 9 fill in details relevant to the conclusion. Only two of the nine references were more than five years old at the time of publication. Two were not in print at the time of the article in question; one is cited from a preprint. This is science truly at the cutting edge of a research front.

Kroto et al.'s reference list follows a consistent style dictated by the journal *Nature*. Conventions of typeface, abbreviation, and punctuation in reference lists vary somewhat from journal to journal and within a journal over time. Some include titles of article, others do not. Some place the volume number in bold, others italicize it. Some give the year of publication in parentheses, others precede it with a comma. All follow the general rule that the various parts of the citation be visually distinct. You should always consult the journal's style guide before finalizing a list. Seeing well-chosen references typed in the correct style gives referees and editors some confidence that the authors have put some thought into assembling their list and writing their article.

Giving Credit by Acknowledgment

The acknowledgment supplements both the list of authors and the citations. It credits those organizations that provided support and those people who were involved in the creation of new science but not directly as scientists. This list typically includes technicians who helped carry out experiments and other scientists who donated materials, were consulted during the course of the research, or critically reviewed a draft of the manuscript. And it is considered a major faux pas to leave out funding agencies.

We excerpt the acknowledgments from a *Nature* paper describing a fossil reported to be the missing link between fish and tetrapod, discovered in the Canadian Arctic. The article itself has only three authors (Shubin, Daeschler, and Jenkins 2006) but a host of other contributors:

> The illustrations are the work of K. Monoyios. Specimen preparation was performed by C. F. Mullison and B. Masek. Permits to conduct this research were granted by the Nunavut Ministry of Culture, Languages, Elders and Youth (D. Stenton and J. Ross). . . . A. Embry and U. Mayr provided guidance at the inception of the field project. M. Coates commented

> on a draft of the manuscript. Field assistance (1999–2004) was provided by W. Amaral, B. Atagootak, J. Conrad, M. Davis [plus nine others]. . . . This research was supported by a patron of our research, the Academy of Natural Sciences, The Putnam Expeditionary Fund (Harvard University), the University of Chicago, the National Science Foundation. . . . Author Contributions: N.H.S. and E.B.D. conceived and co-directed the project. F.A.J. Jr collaborated on all phases of the research.

Note that at the end the above passage specifies each author's contribution to the project. Spelling out such matters is a fairly recent development, still far from commonplace but gaining in popularity. It should.

Deciding whom to acknowledge is usually straightforward, though even that can turn contentious. For example, in the article in question, those providing guidance at the project's start and others who provided field assistance did not qualify as authors. Do you think it would have been legitimate for the three authors to add these individuals to their list? Our next exercise gives you some practice in thinking about who belongs in the author list and who in the acknowledgments.

EXERCISES

Exercise 1

Do you think the following individuals should be listed as authors or appear in the acknowledgments? Why?

1. a chemist who analyzed all of the samples in an experiment by means of techniques she is one of the few people in the world to have mastered
2. a technician who helped in executing all phases in an experiment but was not involved in designing it or analyzing the results
3. a colleague who donated key materials or equipment to an experiment
4. a professor in a science department who raised the funds for a research project carried out by her students and monitored their work but did not actively participate
5. a scientist who thought up an innovative idea for a project and then passed it on for someone else to execute
6. a theorist who analyzed the experimental results and helped formulate the major claims but was not involved in the actual experiment
7. a computer programmer who wrote an innovative program that ran some complex experiment in a particle accelerator
8. a medical writer who wrote the first draft of a clinical medicine paper but did none of the research

Answer

Practices differ among research teams, but in general, the custom today is to include as an author those who have contributed to some aspect of the research project *and* whose career depends upon publication. Mention in the acknowledgments is not generally helpful in career advancement or reputation building. Those who work in support positions and are not expected to publish thus tend to wind up in the acknowledgments. In our list above, that would likely include 1, 2, 7, and 8. Also, we believe no. 3 should appear in the acknowledgments since most scientists do not consider the simple gift of materials or loaning of equipment, no matter how important, a sufficient contribution to warrant authorship. No. 6 almost certainly would have been included as an author. Numbers 4 and 5 could have gone either way.

Exercise 2

In 1986, a group of six biochemists published an article in *Cell* on how immune-specific genes rearrange their DNA to form different antibodies that destroy hostile invaders in the body (Weaver et al. 1986). Trouble began when a postdoctoral fellow (Margot O'Toole) ran into serious problems in reproducing the original experimental work and questioned whether one of the authors (Thereza Imanishi-Kari) actually did the experiments and got the results she claimed. Another author, a Nobel Prize–winning scientist, David Baltimore, had analyzed the results but left the experimenting up to Imanishi-Kari. Because of his involvement, this incident will forever be linked with his name.

After years of intense scrutiny by different investigating committees, no fraud was discovered, but the investigation did find that Imanishi-Kari was guilty of careless housekeeping in conducting her experiments. More important, a few of her statements in the published paper did not jibe with what was actually done, though they did not appear to invalidate the major knowledge claims.

Consider the following questions:

1. Should Baltimore have been held accountable for not better monitoring the work of Imanishi-Kari?
2. Should Imanishi-Kari have been dismissed from her university job for publishing what appear to have been false statements, however minor, about her work?
3. Should the published paper have been retracted even though the authors still firmly believed in the knowledge claims?

Both Baltimore and Imanishi-Kari suffered many indignities and hardships as a result of several investigations spanning nearly a decade. Most important, they spent many hours defending themselves before committees and the press instead of doing science. All but two of the authors did retract their paper, pending the final outcome of the various investigations. If you are curious about more details on what actually happened, consult Daniel J. Kevles's brilliant and engrossing *The Baltimore Case: A Trial of Politics, Science, and Character* (1998).

CHECKLIST

Here is our credit checklist, which may help avoid embarrassing problems after article publication:

- Have all who deserve the designation of authors, that is, all scientists who have contributed directly to the creation of new science, been acknowledged as such?
- Does their order of presentation conform to a rationale that has been spelled out, one that takes into consideration the degree and importance of their participation?
- Have all those who contributed to your research in a meaningful way but have not contributed directly to the creation of new science been acknowledged? Do the acknowledgments include funding agencies?
- Does the list of citations reflect a thorough search for all the relevant scientific literature?
- As an author, are you willing to take responsibility for your results and conclusions should the article you have published become the object of scrutiny because of an accusation of mismanagement or fraud or conflict of interest?

8 Arranging Matters

Picking up a research article for the first time, most scientist-readers can zero in on what information interests them, thanks to a standard arrangement of the elements we have described in the previous chapters, an arrangement often abetted by a system of headings that mark the boundaries for each element. So far we have described those elements in the order in which an author or research group might elect to compose them. In printed form, however, the elements normally follow the arrangement below:

- a *title* that compactly conveys the gist of the main new knowledge claim or claims, followed by a list of authors and their institutional affiliations
- an *abstract* that expands on the title—a brief digest of the important claims and methods
- an *introduction* that places readers in the scientific context in which its authors are working and defines a specific research problem
- a *methods and materials* section that details the steps followed to solve the problem and explain choices behind methods and materials, as necessary for the reader to fully understand the significance of the results that follow
- *results* that display the data generated by the methods, often combined with a *discussion* that interprets and qualifies the data
- a *conclusion* that reiterates the central new claims and addresses future research that would extend the present insights
- *references* that identify sources the authors have relied on, contradicted, or extended
- *acknowledgments* that note personal or financial assistance provided for the article's research

In this chapter, we summarize the lessons of the previous chapters by presenting and commenting upon examples of the main substantive sections in the typical scientific article, ordered to conform to the standard arrangement. Our examples draw upon diverse sciences: astrophysics, genetics, stereo-chemistry, experimental biology, sociology, and cultural anthropology. In the next chapter, we examine the interlocking parts of several whole articles from beginning to end. We hope these two chapters will give our readers a better sense of how to assemble the standard parts into a unified whole and also how to vary the standard parts to meet the specific needs of different kinds of research articles.

Title and Abstract

In 1995, Michel Mayor and Didier Queloz published an article that provided firm evidence for the existence of a planet outside our solar system. Their title, "A Jupiter-Mass Companion to a Solar-Type Star," is like a headline: no word is wasted in focusing on this breaking news. The nucleus noun *companion* emphasizes their newly discovered astronomical body. Its modifiers fore and aft provide necessary specifics: readers infer that this companion is a planet like Jupiter circling a star like our sun.

The equally succinct abstract outlines the authors' conclusions, establishes the means by which these conclusions were derived—inference rather than direct observation—and, finally, airs two plausible explanations of planetary origin, grounded in current astronomical theory:

> The presence of a Jupiter-mass companion to the star 51 Pegasi is inferred from observations of periodic variations in the star's radial velocity. The companion lies only about eight million kilometers from the star, which would be well inside the orbit of Mercury in our Solar System. This object might be a gas-giant planet that has migrated to this location through orbital evolution, or from the radiative stripping of a brown dwarf.

The abstract's first sentence amplifies the title, an effective and common opening strategy for an abstract. Note three other exemplary qualities. Within a mere three sentences this abstract tells us what the authors did (inferred the existence of an extrasolar planet), how they did it (astronomical observations of the trajectory and speed of a solarlike star), and what they discovered as a result (the planet's distance from the star, its type, and its possible history).

Yet another exemplary literary quality is that this abstract's sentences are short and their syntax is simple. Nevertheless, full understanding is not immediate except for an astrophysicist. Understanding this passage depends on acquisition of the knowledge and vocabulary of a specific field: "51 Pegasi," "radial velocity," "orbital evolution," "radiative stripping," "brown dwarf." We will be discussing this aspect of scientific writing in chapter 14.

Introduction

In 1997, Ian Wilmut and his collaborators published the startling news that they had cloned a viable mammal, a sheep named Dolly. In the more technical language of their abstract, "The fact that a lamb derived from an adult cell confirms that differentiation of that cell did not involve irreversible modification of genetic material required for development to term."

Scientific introductions rarely attempt to engage the reader with an opening anecdote or startling fact. Their most important job is to define a problem within a research territory and, in so doing, convince knowledgeable readers it really is a problem worth solving. In the five sentences of their introduction, as in a film sequence, Wilmut and his coauthors move from a long shot of the state of the field to a narrow focus on a particular and difficult problem at its cutting edge: the transfer of a single nucleus in an adult cell to an unfertilized egg without genetic damage during cell differentiation. The rationale for this ordering is clear: to place the authors' current work in the context of research in their field and, as a consequence, to open up a space for their own efforts.

> It has long been known that in amphibians, nuclei transferred from adult keratinocytes established in culture support development to the juvenile tadpole stage. Although this involves differentiation into complex tissues and organs, no development to the adult stage was reported, leaving open the question of whether a differentiated adult nucleus can be fully reprogrammed. Previously we reported the birth of live lambs after nuclear transfer from cultured embryonic cells that had been induced into quiescence. We suggested that inducing the donor cell to exit the growth phase causes changes in chromatin structure that facilitate reprogramming of gene expression and that development would be normal if nuclei are used from a variety of differentiated donor cells in similar regimes. Here we investigate whether normal development to term is possible when donor cells derived from fetal or adult tissue are induced to exit the growth cycle and enter the G0 phase of the cell cycle before nuclear transfer.

As we move from the field as a whole to the authors' contribution to it, the introduction shifts legitimately from the impersonal to the personal. The shift is from *it* and *this* to *we* as the grammatical subject. It is a shift away from the body of accumulating knowledge and onto the authors themselves as actors in a drama of considerable intellectual importance, one that is also being played for high ethical stakes.

Methods

Although sections describing the means by which the authors solved their problem may be the least read of all in published articles, they will be carefully read by manuscript reviewers: a common reason given for manuscript rejection is that its methods section has serious flaws. Of course, the methods section can become the center of attention for readers in fields where methodological breakthroughs are routine or where the methods are particularly innovative. Such was the case in 1958 when John Kendrew and his collaborators sought to determine the structure of a complex protein (myoglobin)—the sequence of amino-acid building blocks along its chain, the regular conformations along the backbone of this chain, and the relative positions of the atoms of both the backbone and side chains. For that purpose they employed the then relatively new technique of x-ray diffraction. To give some perception of the problem they faced, we should note that one million trillion proteins would fit on the head of a pin. First, the authors purified the protein and crystallized it, then they passed x-rays through it, and then, on the basis of the reflections so produced, they reconstructed the protein by mathematical techniques.

We excerpt the first paragraph from a section titled "Methods of X-ray Analysis." Even to general readers, those who like ourselves are soon lost in its intricacies, this exposition forcefully conveys the fiendish difficulty of reconstructing the complex protein molecule in the x-, y-, and z-dimensions, the three dimensions of Euclidian space. The structure of the paragraph is clear and clearly procedural. Step by step, Kendrew and his coworkers go about their highly technical business. Still, we must not be misled by this step-by-step approach; we must not think that scientific papers are in any legitimate sense the equivalent of a cooking recipe. What is related is not a recipe but an integral part of an argument claiming that the authors have solved the structure of myoglobin:

> Type *A* crystals of myoglobin [protein in tissue that receives oxygen from hemoglobin and stores it until needed] are monoclinic (space group $P2_1$) and contain two protein molecules per unit cell. Only the *hol* reflexions [resulting from x-ray diffraction] are "real," that is, can be regarded as having relative phase angles limited to 0 or π, or positive or negative signs, rather than general phases; when introduced into a Fourier synthesis, these reflexions give a projection of the contents of the cell along its y-axis. In two dimensions the analysis followed lines similar to that of hæmoglobin

[protein in blood that carries oxygen to tissue]. First, the heavy atom was located by carrying out a so-called difference-Patterson synthesis; if all the heavy atoms are located at the same site on every molecule in the crystal, this synthesis will contain only one peak, from the position of which the *x*- and *z*-co-ordinates of the heavy atom can be deduced and the signs of the *hol* reflexions determined. These signs were cross-checked by repeating the analysis for each separate isomorphous replacement in turn; we are sure of almost all of them to a resolution of 4 A. [angstrom, equal to 0.1 nanometer], and of most to 1.9 A. Using the signs together with the measured amplitudes, we may, finally, compute an electron-density projection of the contents of the unit cell along *y*; but as in hæmoglobin and for the same reasons, the projection is in most respects uninterpretable (even though here the axis of projection is only 31A.). On the other hand, knowledge of the signs of the *hol* reflexions to high resolution enabled us to determine the *x*- and *z*-co-ordinates of all the heavy atoms with some precision. This was the starting point for the three-dimensional analysis now to be described.

Results

In 1976, Erwin Neher and Bert Sakmann perfected a method for measuring the flow of electricity in animal membranes, the "patch clamp" method. Although a new method is at the very center of the discovery, Neher and Sakmann still needed to make the case that their method worked. For that they needed to test it, report the results, and explain the meaning of those results:

> Figure 2 shows current recording taken in the conditions outlined above. Current can be seen to switch repeatedly between different levels. The discrete changes are interpreted as the result of opening and closing of individual channels. This interpretation is based on the very close similarity to single-channel recordings obtained in artificial membrane systems. The preparation under study is, however, subject to a number of different sources of artifact. Therefore it is necessary to prove that the recorded events do show the properties which are assigned to ionic channels of the cholinergic system. These are: a correlation with the degree of hypersensitivity of the muscle membrane; an amplitude dependent on membrane potential as predicted by noise analysis; a mean length or channel open time, which should depend on voltage in a characteristic manner; pharmacological specificity with different mean open times for different cholinergic agonists. The experiments bore out all of the above-mentioned points as outlined below.

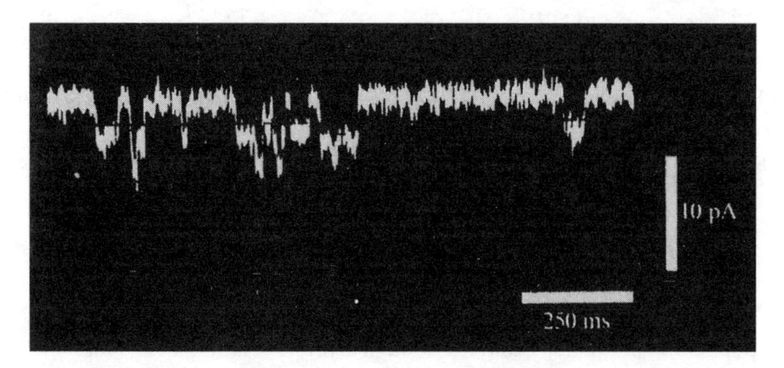

Oscilloscope recording of current through a patch of membrane of approximately 10 μm^2 . Downward deflection of the trace represents inward current. The pipette contained 2×10^{-7} M SubCh in Ringer's solution. The experiment was carried out with a denervated hypersensitive frog cutaneous pectoris (*Rana pipiens*) muscle in normal frog Ringer's solution. The record was filtered at a bandwidth of 200 Hz.

FIGURE 4. The "patch clamp" method in action: An oscilloscope reading. Reprinted with permission of Macmillan Publisher Ltd. (Neher and Sakmann 1976).

This selection is typical in that its meaning is dependent not only on words but on a visual, in this case, an oscilloscope recording of current (*y*-axis in units of picoamperes) as a function of time (*x*-axis in units of milliseconds; see our fig. 4, which is fig. 2 in Neher and Sakmann). Once interpreted, the figure counts as evidence of the fluctuation of electrical currents in animal tissue, in this case the opening and closing of channels. The downward deflection represents inward current. There is a genuine division of labor in this excerpt: the figure reveals the phenomenon in question, the figure legend explains it in detail. The text also links the figure and the detail to the general claim, presented later in the article, that "the observed conductance changes are indeed recordings of single-channel currents." Put simply, the results demonstrate that the method works.

Discussion

While results sections are not free from argument, and while the whole of the scientific paper constitutes an argument, inference to the best explanation is at the very center of discussion sections. We see this centrality operating in a 2001 article by Jody VanLaningham, David Johnson, and Paul Amato, reporting their discovery that the U-shaped curve routinely thought to track marital happiness—highs at both ends of long marriages with a long, deep valley in between—is in fact an artifact of defective methodologies. Instead of ups and downs, there is (alas!) a general and steady decline. In this passage of their discussion, these scientists support the conjecture that the U-shaped artifact is a consequence of a difference between generational expectations:

> Regardless of the explanation for the long-term decline in marital happiness, our results demonstrate that the U-shaped pattern observed in most cross-sectional studies is artifactual. The U-shaped association apparent in cross-sectional data could be due to the gradual departure from the married population of unhappy couples through divorce, resulting in an "increase" in mean happiness among couples in long-term marriages. But this explanation is unlikely, given that most divorces occur early in marriage and the observed rise in happiness does not appear until after twenty years of married life. We believe that the cohort explanation advanced by Glenn (1998) is a more likely contender. Glenn's analysis specifically ruled out the possibility that attrition through divorce resulted in compositional differences across cohorts. Instead, his analysis suggested that the apparent U-shaped association between marital duration and marital happiness is due to older marriage cohorts experiencing higher levels of marital happiness than younger marriage cohorts. These older cohorts—married at a time when people held more pragmatic views about marriage, support for marriage was stronger, and couples were more committed to the norm of lifelong marriage—may have strengths that allow them to maintain high levels of marital happiness.

Conclusion

While introductions and results focus on the past (what has been discovered) and discussions focus on the present (what these discoveries mean now), conclusions often look toward the future. In conclusions, scientists speculate on the significance of the present research in which they are engaged and on possible directions of future work. The article from which the conclusion below is excerpted is by Francesco d'Errico and his associates (2005). Its subject is the discovery of shell beads in the Blombos Cave in South Africa, from whose presence these researchers infer that complex societal behaviors characteristic of human modernity arose earlier in African prehistory than had been conjectured previously. In their conclusion, d'Errico and his associates summarize the objections to this hypothesis and state clearly how their new evidence obviates these. Of course, as they point out at the very end, this is only another step forward in a long scientific journey, one that might one day result in a complete picture of the "emergence of modern humanity." While their grammar is a bit convoluted at points—an artifact, perhaps, of writing outside their native language—their reasoning is not:

> Evidence for an early origin of modern human behaviour in Africa has long remained elusive. Recent finds in > 70 ka [thousands of years] African sites of objects bearing abstract engravings, large quantities of pigment and formal bone tools (Yellen *et al.*, 1995; McBrearty & Brooks, 2000; Henshilwood *et al.*, 2001a, b, 2002) have been rejected as clear-cut evidence for behavioural modernity on the ground that context, dating, and/or because deliberate symbolic intent could not be warranted (Wadley, 2001, 2003; Klein, 2000; Ambrose, 2001). The discovery of personal ornaments in the c. 75 ka MSA [Middle Stone Age] layers at BBC [the Blombos Cave] adds an unambiguous marker of symbolically mediated behaviour to the list of innovations already identified in the MSA.
>
> Since syntactical language is the only means of communication bearing a built-in meta-language that permits creation and transmission of other symbolic codes (Aiello, 1998), beadwork represents a reliable proxy for the acquisition of language and fully modern cognitive abilities by southern African populations 75,000 years ago.
>
> . . . The BBC beads clearly predate the arrival of AMH [Anatomically Modern Humans] in Europe and the 50,000 years old rapid neural mutation that would have qualitatively changed, according to some authors, human cognition. Since personal ornaments cannot be considered the only hallmark of modernity, are not the only means human cultures use for body decoration, and are often made out of perishable raw material, we can hardly deny modernity to contemporary Neanderthals on these grounds nor rule out that *H. sapiens* were behaviourly modern before 75 ka. Neanderthals show cultural innovations such as burials, pigment use and, at a later stage, personal ornaments suggesting their ability to create symbolic cultures.
>
> Future research needs to establish a geography and precise chronology for behavioural innovations in Africa and Eurasia with the aim of understanding the role played by each in the emergence of modern humanity.

Our Conclusion

From these excerpts, we hope you can see that good scientific articles combine rigorous argument and ritual observance. Each of these aspects serves science: rigorous argument, founded directly or indirectly in empirical data, convinces scientists of the truth of claims; ritual observance of routine principles of organization makes complex scientific arguments easier to follow.

But be forewarned that good scientific writing is not simply a matter of mechanically following a blueprint like that beginning this chapter. That blueprint is meant only as a starting point from which you can creatively erect an argument that will make the strongest possible case for a proposed knowledge claim. Different subject matter and even the same subject matter for different audiences may dictate a different structure and content. That is the subject of our whole next chapter.

EXERCISE

Search the Web site of a scientific journal in an area of interest to you. Open a long article that has a network of headings and subheadings. Reproduce those on a separate sheet of paper or Word document. Here is an example that caught our attention from a recent issue of the *Journal of the American Medical Association* (Bach et al. 2007):

TITLE: "Computed Tomography Screening and Lung Cancer Outcomes"
INTRODUCTION
METHODS
  Prediction Models
  Modifications of Prediction Models
  Outcomes
  Clinical End Points
  Mortality and Survival End Points
  Statistical Analysis
RESULTS
  Characteristics of Studies
  Frequency of Lung Cancer Diagnosis and Lung Cancer Resection
  Number of Advanced Lung Cancers and Deaths Due to Lung Cancer
  Relationship between Initial Lung Cancer Diagnosis and Death Due to Lung Cancer
COMMENT
AUTHOR INFORMATION
  Corresponding Author
  Author Contributions
  Financial Disclosures
  Role of Sponsors
  Acknowledgment
  Author Affiliations
REFERENCES

Now compare your headings against the standard arrangement of elements and the content of those elements as we defined them in the earlier chapters. What elements conform? How do they deviate? You might want to compare your article's contents against the checklist that ends this chapter. In the case of our selected medical article, you will find our answer in the next chapter.

CHECKLIST

Whether you are submitting your first or hundredth research article for publication, asking the following questions will help make it a worthy contribution to the scientific literature:

- Does the title reflect the claim the article is making, with the key aspect of that claim in the head noun?
- Does the byline list only authors deserving of author credit? Has anyone been omitted or given honorary authorship? Is there some reason behind the order in which authors are mentioned?
- Does the abstract capture the main claim and method used? Would a sentence or two setting the context help the reader better appreciate the significance of the claim?
- Does the introduction establish a research problem and contextualize it within a research territory to an extent appropriate for the intended audience? Will the reader understand why solving the problem is important?
- Is the method description complete enough so that others in the field can judge whether it represents a plausible strategy for solving the problem mentioned in the introduction? Does the methods section recount the key preparations needed to conduct the research and the means necessary to analyze its results?
- Do the results appear credible based on the methods used to generate them?
- Are the results represented in figures and tables in a way that supports any claims derived from them? Does the accompanying text adequately explain the meaning of those figures and tables?
- Does the discussion establish original knowledge claims based on the results and make a sound argument for those claims, including any qualifications or alternative explanations?
- Does the conclusion reiterate knowledge claims, place those claims within a wider context, and suggest future lines of inquiry?
- Does the references list credit all of the key works for the research, published and unpublished?
- Do the acknowledgments thank those who are not authors but who nevertheless contributed to the research project? Are funding sources included?
- If the article has different authors responsible for different aspects of the research, are the contributions of each author specified in a separate section?

9 Varying Matters

The standard arrangement discussed in the previous chapter applies primarily to experimental articles, that is, those recounting the manipulations of natural and manufactured objects using a prescribed method that solves a research problem by generating and analyzing data. Such articles, which provide an empirical basis for the continued conceptual evolution of science, dominate the pages of the twentieth- and early twenty-first-century journal literature. But different content can dictate a moderate or even substantial variation on the standard arrangement. In this chapter, we guide you through a close reading of successful research articles of three different types: clinical medicine, theory, and review of literature. We compare each type to the template given in the previous chapter.

Clinical medicine articles report on tests of some new diagnostic method or treatment on a group of patients or volunteers. They can have profound repercussions on our health care. We discuss an article that assesses the medical advantages of routine computerized x-ray scanning for individuals at risk for lung cancer (Bach et al. 2007).

Theoretical articles explain natural events in a new way, often suggesting experiments or observations that might support the explanations. They provide the conceptual breakthroughs that drive the continued evolution of science. The sample theoretical article we selected represents a conceptual breakthrough in the field of evolutionary biology, a radical new system for classifying species (Woese, Kandler, and Wheelis 1990).

Review articles describe and evaluate the recent published literature in a field—usually limited to the previous decade. Unlike the standard article reporting original results, reviews are the product of trips to the library and office and Web, rather than to the laboratory or field site. They summarize and interpret past science in order to provide a synoptic view that will help other scientists acquaint themselves with unfamiliar territory. In the best research journals like *Physical Review, Nature,* and *Science* they also provide a new perspective on familiar territory for those working in the same area, as does our selected sample, which reviews an advanced research technique used in cognitive neuroscience, functional magnetic resonance imaging, better known by the shorthand fMRI (Logothetis 2008). We also examine the inner workings of a hybrid article—part literature review, part theoretical article—that proposed a new model for the regulatory mechanism of genes (Jacob and Monod 1961).

We close this chapter with an experimental-type digital article whose arrangement differs substantially from the standard one because it takes advantage of new organizational strategies possible due to the World Wide Web (Green et al. 2006).

Medical Article

In contrast to the typical scientific article in specialized journals, the audience for a medical article might include many who are not doing research in the field. In particular, the audience might include physicians interested in applying the reported information to their patients. It might also include citizens with a vested interest in the contents: either they or a loved one might have the medical condition under scrutiny. Yet another community of potential readers is makers of health care policy in government and industry. With such a diverse readership, establishing the societal implications of the research in the introduction or conclusion (or both) tends to be very important.

The components of the typical research article in clinical medicine conform to those given in the previous chapter but with some important variations in content:

- a title touching upon a possible intervention in the diagnosis or treatment of a current medical problem
- an abstract summarizing an unproven intervention for a medical problem, the design of the trial employed to assess it, and the outcome from the trial
- an introduction that reviews current medical practice with regard to the diagnosis or treatment of a disease, identifies a problem with that practice, and notes the intervention that the authors believe might improve it
- a methods section describing the authors' trial of the proposed intervention with a representative sample of volunteers; typically, that entails describing what are the aggregate characteristics of the individuals participating in the trial (age, sex, number, length of participation, medical history, etc.), what was done to them relevant to the medical problem posed in the introduction, and how data about the participants was gathered and statistically analyzed; the methods section holds considerable importance

in clinical medicine articles because their credibility largely rests on the research project's design, not the accuracy of any measuring equipment
- results from having applied the intervention to the participants described in the methods section, comparison with what happens without the intervention, and a statistically based analysis of whether the intervention makes a significant difference
- conclusions about the effectiveness of the intervention with the trial participants, the relevance to current medical practice and policy, and future possible research
- references, acknowledgments, and statement on any possible financial conflicts of interest

We chose for analysis a clinical medicine article asking whether annual screening of the lungs with computed tomography (commonly abbreviated as "CT") offers any benefit for current or former smokers (Bach et al. 2007). Common sense dictates that if you have a powerful imaging technique like CT for detecting small tumors, you ought to use it on a periodic basis with individuals at risk to develop life-threatening tumors. This scientific article concluded that in this case common sense probably does not make sense.

TITLE AND ABSTRACT

The medical article that is our example has the simple title "Computed Tomography Screening and Lung Cancer Outcomes." Its two nucleus nouns announce the twin focus of the article: *screening* of smokers by CT and the *outcomes* of that screening for lung cancer.

Below the title, adjacent to the lists of authors, we find the abstract. It summarizes the article in a way meant to be easily comprehensible to any motivated nonprofessional. In fact, the content of the abstract conforms to a template dictated on the Web site of the *Journal of the American Medical Association* (better known as JAMA), with each element of the template appearing as a heading within the abstract. We reproduce those headings in the column on the left below. The middle column gives our summary of the authors' text accompanying each heading. Each heading on the left conforms to one of our four elements from chapter 2, listed in the column on the far right.

Context	Some smokers are undergoing routine CT screening for lung cancer, but no one has rigorously tested its effectiveness.	*Why is research important?*
Objective	Assess effectiveness of this screening on lung cancer outcomes.	*What was done?*

Design, Setting, and Participants	3246 current or former smokers initially screened for lung cancer were followed for 2–4 years.	*How was it done?*
Intervention	Annual CT scans and treatment of any detected nodules	*How was it done?*
Main Outcome Measures	Data on new lung cancers detected, medical interventions that result, and deaths related to the disease	*What was discovered?*
Results	Three times more cases of lung cancer detected, but no statistical evidence of decline in number of diagnosed cases of advanced lung cancer or deaths related to lung cancer	*What was discovered?*
Conclusions	Annual CT screening for lung cancer may not be beneficial	*Why is discovery important?*

INTRODUCTION

The article introduction conforms to the standard structure we outlined in chapter 1. The authors begin by broaching a societal problem: reciting a few chilling statistics about lung cancer deaths. They then formulate a premise that, at first glance, would appear to be obvious: "Screening individuals at high risk for lung cancer might reduce these statistics." But, they note, this premise proved not to be the case in earlier studies of chest x-ray screening. The medical diagnosis problem they pose is, how do we know the same is not true of CT screening? In general, clinical medicine articles arise out of some problem regarding the diagnosis and treatment of a disease or medical abnormality, not necessarily past research by others.

METHODS

The methods section appears in the standard order but does *not* adhere to the chronological structure we proposed in chapter 6. Instead, it consists of three components related to different aspects of a research project involving current or former smokers who volunteered to participate in a trial. The first has to do with measuring the outcomes from routine CT lung scanning performed at three medical institutes (Instituto Tumori in Italy, Mayo Clinic, and Moffitt Cancer Center) over several years. The second concerns the mathematical models the authors used to predict lung cancer outcomes that would have occurred in the absence of CT scans. The third describes the statistical methods used for analysis of the measured and predicted results. From this section, we learn that the overall argument hinges on a comparison of clinical outcomes and model predictions combined with statistical analysis.

RESULTS

The results section opens with relevant statistics regarding the individuals participating in the three studies (age, sex, smoking history, number of participants screened at each medical center, and years followed in the study). Those characteristics could have just as easily appeared in the methods section. But they appear at this juncture so that the authors can insert a table that combines them with the outcomes related to CT scanning: number of new lung cancer cases, resulting lung surgeries, and deaths. This table (reproduced here as fig. 5) has four columns of data for ease of comparison: one for each of the three studies and one for the total. It also has two halves. The top half—from "Age" to "Year since quitting for those who had quit"—concerns the medical history about the participants; the bottom half—beginning with

Table 1. Characteristics of Studies Included in Analysis of Computed Tomography Screening for Lung Cancer*

	Total (N = 3246)	Istituto Tumori (n = 977)	Mayo Clinic (n = 1439)	Moffitt Cancer Center (n = 830)
Age, mean (SD), y	60.1 (6.4)	58.4 (5.4)	60.0 (6.3)	62.2 (7.0)
Sex, No. (%)				
Male	1917 (59)	693 (71)	737 (51)	487 (59)
Female	1329 (41)	284 (29)	702 (49)	343 (41)
Smoking history				
Years of smoking, mean (SD)	39.1 (7.2)	38.0 (6.4)	39.2 (7.2)	40.4 (8.0)
Cigarettes smoked per day, mean (SD)	27.1 (10.2)	25.2 (10.0)	26.7 (9.9)	30.1 (10.3)
Quit at study entry, No. (%)	1043 (32)	121 (12)	551 (38)	371 (45)
Years since quitting for those who had quit, mean (SD)	7.1 (5.0)	7.3 (6.0)	5.1 (3.2)	10.0 (5.5)
Clinical end point†				
Follow-up to diagnostic outcomes				
Person-years	10 736	3482	5199	2055
Median, y	3.9	3.9	4.0	2.1
Diagnosed with lung cancer				
No. of persons	144	36	66	42
Rate per 1000 person-years	13.9	10.3	12.7	20.4
Diagnosed with advanced lung cancer				
No. of persons	42	11	18	13
Rate per 1000 person-years	3.9	3.6	3.5	6.3
Follow-up to surgical outcomes				
Person-years	10 738	3484	5200	2055
Median, y	3.9	3.9	4.0	2.1
Lung cancer resection				
No. of persons	109	33	48	28
Rate per 1000 person-years	10.2	9.5	9.2	13.6
Mortality end point	(N = 3210)	(n = 976)	(n = 1432)	(n = 802)
Follow-up from 1 year to death or censoring				
Person-years	10 942	2592	6217	2133
Median, y	3.7	3.0	4.4	2.2
Lung cancer deaths				
No. of deaths	38	7	19	12
Rate per 1000 person-years	3.5	2.7	3.1	5.6

*The entry dates for the studies are as follows: Istituto Tumori (September 9, 2000, through July 13, 2001), Mayo Clinic (January 18, 1999, through December 17, 1999), and the Moffitt Cancer Center (December 11, 1998, through August 13, 2003).
†First diagnosis of lung cancer and first surgery.

956 JAMA, March 7, 2007—Vol 297, No. 9 (Reprinted)

FIGURE 5. Patient characteristics and outcomes for clinical medicine article. Reproduced with permission of American Medical Association (Bach et al. 2007).

"Clinical end point"—relates outcomes. Note the two major headings in the bottom half—"Clinical end point" and "Mortality end point"—correspond to subheadings in the methods section. The remainder of this section discusses the clinical outcomes in the table, their comparison with the model predictions, and the statistical significance of the key comparisons—typical content for a results section, as we discussed in chapter 4.

COMMENT

What the authors call the "comment" section is equivalent to what we referred to as "discussion" in chapter 4. Here we learn that, statistically speaking, the CT screening significantly increased the number of cases of lung cancer detected, as we would expect, but somewhat unexpectedly, it did not reduce the number of cases diagnosed with *advanced* lung cancer, nor did it reduce the number of deaths from lung cancer. Hence CT scanning had no apparent benefit. Worse, it resulted in biopsies and lung resections for small tumors that did not statistically alter the final outcome. The authors bolster their present findings by favorable comparisons to results from earlier cancer-screening tests.

They close their discussion section with a penultimate paragraph on their study's limitations and a final paragraph on the relevance of their study to medical practice and legislation. While not given a "conclusion" heading, these last paragraphs serve that purpose. We quote the final paragraph in full because it touches upon all three elements of the scientific conclusion as defined in chapter 5. The final sentence makes for a powerful closing:

> Despite the paucity of evidence supporting lung cancer screening, and no clear delineation of the harms that may result from excess diagnoses, additional diagnostic procedures, and additional treatment, screening is being offered widely, and claims that screening saves lives and should be available to all are widespread.[25, 46–47] Legislation has also been introduced that would require Medicare to cover lung cancer screening.[48] A more prudent course would be to await the findings of the National Lung Screening Trial and several trials that are being conducted and planned in Europe. It would also be wise to explore other approaches to lung cancer prevention and early detection based on modalities other than regular imaging. Until then, CT screening for lung cancer should be considered an experimental procedure, based on an uncorroborated premise.

In short, despite some promising results published by others, the authors' position is that the jury is out on the benefit of routine CT screening for lung cancer, and that conclusion remains true today. (For an excellent status report as of 2008, see Mulshine.)

AUTHOR INFORMATION

The very end of this article has the usual acknowledgments and references. Typical for JAMA articles but not for scientific articles in general, it also specifies who on the research team did what in terms of research concept and design, collection of data, writing and revision of the article, statistical analysis, acquisition of needed research funds, and supervision. The practice of assigning responsibility for different aspects of the research has been instituted by some journals in response to several well-publicized cases of fraudulent statements by certain individuals on research teams. Some authors have taken to incorporating this sort of information in their acknowledgments even when the journal style does not request it. We would not be surprised to find this practice becoming routine in the near future.

The author information section also includes statements concerning any financial interest either the authors or their sponsors might have in the outcome. This is not standard practice for scientific articles in many disciplines. But unlike the typical scientific article, medical articles can have a major impact on industrial and governmental practices, as well as the lives of ordinary citizens. This information protects against bias due to conflicts of interest.

Theoretical Article

The structure of the theoretical article is not nearly as firmly entrenched as that of its experimental counterpart. However, it typically includes some or all of the following elements:

- a title that captures in abbreviated form the authors' new or modified theory for the working of nature
- an abstract that recounts the authors' theory and the main evidence or argument supporting that theory
- an introduction that identifies a conceptual problem within existing theory or points to a conflict between existing theory and experimental results and also gives the author's approach to solving the problem
- a main body with two components: first, a theoretical solution deduced from various assumptions, experiments and observations, definitions, and boundary conditions; second, a proof that arises from a comparison of theoretical predictions with experimental results or observations
- a conclusion that reiterates the essential points of the new or modified theory and proposes future work that would verify or extend it
- references and acknowledgments as in the experimental type

Our example is a short theoretical article proposing a new taxonomic system for the known living organisms (Woese, Kandler, and Wheelis 1990). In it the authors propose reducing the reigning taxonomy from five "kingdoms" to three "domains."

TITLE AND ABSTRACT

The article title has two components separated by a colon: "Towards a Natural System of Organisms: Proposal for the Domains Archaea, Bacteria, and Eucarya." The main component places the emphasis on the nucleus noun *system,* while the subordinate component stresses *proposal.* This structure makes the authors' purpose clear: they are *proposing* a new *system.* The first word in the title is the preposition *towards;* it emphasizes the tentative nature of the authors' proposal. That tentativeness is in keeping with a claim the authors know will prove controversial.

The abstract does not follow the conventions we described at the start of chapter 2. It begins with a *generalization:* biomolecular information is "more revealing of evolutionary relationships" than observed morphological relationships or the fossil record. A *consequence* follows: the evidential basis for evolutionary biology has shifted over the past century from multicellular organisms to single cells to genetic molecules. A *proposition* is implicit: past evolutionary trees are flawed and ought to be replaced by the authors' new one based on DNA and RNA analyses at a molecular level. The authors are not proposing a revision of the old system but an entirely new system resting on firmer empirical support. The abstract summarizes the authors' argument for that system.

INTRODUCTION

The authors give this section a name that captures the essence of their research problem—"Need for Restructuring Systematics"; nonetheless, the organization of their introduction otherwise conforms to the standard format defined in chapter 1. It poses a problem and implies a solution grounded in three observations concerning past taxonomic practices. The first is that, as a consequence of great advances in analyzing genetic molecules, "it has become possible to trace evolutionary history back to" nearly the beginning of microbial life on earth. Before these advances, knowledge about microbial life was restricted to only the past 20 percent of life on earth, a limit that did not take into account the vast tract of time when microorganisms predominated. The second observation is that for many centuries scientists based the tree of life on the "ancient notion that all living things are either plant or animal," a mistaken assumption that had to be discarded before systematics

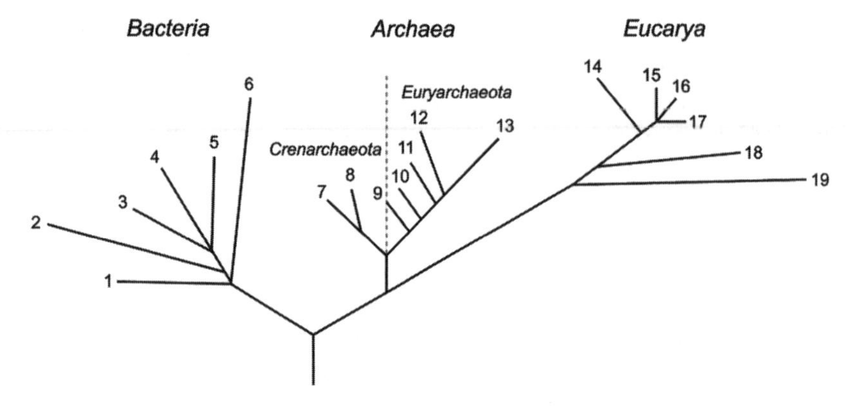

Universal phylogenic tree in rooted form, showing three domains. Branching order and branch lengths are based upon rRNA sequence comparisons. The position of the root was determined by (the few known) sequences of pairs of paralogous genes that diverged from each other before the three primary lineages emerged from their common ancestral condition. This rooting strategy in effect uses the one set of (aboriginally duplicated) genes as an outgroup for the other. The numbers on branch ends correspond to the following organisms:

Bacteria: 1, the Thermotogales; 2, the flavobacteria and relatives; 3, the cyanobacteria; 4, the purple bacteria; 5, the Gram-positive bacteria; and 6, the green nonsulfur bacteria.

Archaea: the kingdom Crenarchaeota: 7, the genus *Pyrodictium;* 8, the genus *Thermoproteus;* 9, the Thermococcales; 10, the Methanococcales; 11, the Methanobacteriales; 12, the Methanomicrobiales; and 13, the extreme halophiles.

Eucarya: 14, the animals; 15, the ciliates; 16, the green plants; 17, the fungi; 18, the flagellates; and 19, the microsporidia.

FIGURE 6. Genetic tree diagram from theoretical article proposing division of living organisms into three domains. Reprinted with permission of author (Woese, Kandler, and Wheelis 1990).

could advance. The third observation is that evolutionary biologists have since accepted three additional kingdoms: single-cell forms, bacteria, and fungi. While these five divisions represent an improvement over the "aboriginal plant/animal division," they are "phylogenetically" flawed, that is, they are not entirely consistent with what DNA sequencing teaches us about the evolution of life on earth.

The authors' solution is to base all evolutionary relationships in the tree of life on data acquired by analysis of the genetic codes among different life forms, using "the classical gross properties of cells and organisms . . . largely to confirm and embellish."

THEORY AND PROOF

The article then turns to the authors' proposal for a "universal phylogenetic tree." This consists of the three domains named in the title: bacteria, archaea (microbes that differ biochemically and genetically from bacteria), and eucarya (plants, animals, fungi, and some single-cell organisms). Branching off from each domain are six or seven kingdoms. The authors portray these relationships not with words alone but with a diagram in which plants and animals are each one small branch on the tree of life (nos. 14 and 16, respectively; see fig. 6). In terms of longevity and evolution, the invisible world of microbes rules.

The core argument rests ultimately on an observation made in the 1960s by Emile Zuckerkandl and Linus Pauling: biomolecules are ideally suited to establishing evolutionary relationships. From that basic premise the authors' new universal tree of life grows. Its support consists of quantitative results involving strands of ribosomal RNA in genetically related organisms.

There is a clear division of labor in this section: the diagram depicts the proposed new theory, its legend explains it in detail, and the main text links the diagram and the detail to the general claim that "molecular structures and sequences are generally more revealing of evolutionary relationships than are classical phenotypes (particularly so among microorganisms)."

CONCLUSION

The conclusion paragraph summarizes the present and future advantages of the proposed "natural system" in the article title.

Review Article

Review articles evaluate the recent literature in a field. In any field alive with activity, the proliferation of published scientific papers of varying quality and importance makes these acts of judgment essential to intellectual advance. Good literature reviews, we find, tend to favor this general arrangement:

- a title and abstract that capture the research territory reviewed and the results of that review; this is a particular challenge because review articles tend to be much longer than those reporting original research, even though the abstracts and titles are of about the same length
- an introduction designed to secure the readers' attention and focus on a central research problem or territory

- a survey of current and past research that summarizes and critically evaluates what is known about the research problem or territory—for example, the authors might identify knowledge gaps that need filling or methodological flaws or inconsistencies in published findings
- a conclusion that does some or all of three things: captures the state of the art in a research territory, formulates some new knowledge claim based on previous literature, suggests future research
- a reference list far surpassing in length that of the typical research article

Moreover, as our next two examples illustrate, the best literature reviews do much more than merely string together summary paragraphs on selected articles or book chapters devoted to a narrow topic; they synthesize previous knowledge claims into something new.

TITLE AND ABSTRACT

Our sample review article is "What We Can Do and What We Cannot Do with fMRI" (Logothetis 2008). As its unorthodox title suggests, this article presents us with a critical evaluation of the literature leading to new knowledge that will be of practical benefit to other researchers, that will tell them "how far fMRI can go in revealing the neuronal mechanisms of behavior." Logothetis's abstract elaborates in straightforward language:

> Functional magnetic resonance imaging (fMRI) is currently the mainstay of neuroimaging in cognitive neuroscience. Advances in scanner technology, image acquisition protocols, experimental design, and analysis methods promise to push forward fMRI from mere cartography to the true study of brain organization. However, fundamental questions concerning the interpretation of fMRI data abound, as the conclusions drawn often ignore the actual limitations of the methodology. Here I give an overview of the current state of fMRI, and draw on neuroimaging and physiological data to present the current understanding of the haemodynamic signals and the constraints they impose on neuroimaging data interpretation.

Note that the abstract follows the pattern from chapter 2 of establishing what was done (survey of fMRI in brain research), what was discovered (the technique's limitations), and why the research is important (identification of these limitations should allow practitioners to better understand data from this "current mainstay of neuroimaging"). The authors do not address how the research was done because it is not that important, as is true for most review-type articles.

INTRODUCTION

The introduction is a much expanded version of the abstract and title. Its first paragraph establishes the importance of the subject matter. Nikos Logothetis notes that MRI is an invention comparable in importance with the x-ray: it "has assumed a role of unparalleled importance in diagnostic medicine and more recently in basic research." Moreover, the advanced technique of fMRI, he asserts, has "had a real impact on basic cognitive neuroscience research" starting in the early 1990s. That statement is supported by noting that in that decade his computer search uncovered "over 19,000 peer-reviewed articles" on fMRI.

The following paragraphs discuss problems with the use of fMRI in cognitive neuroscience: they recount common misunderstandings about what it can and cannot do. Then we get a preview of Logothetis's main conclusion: the fMRI signal does not reflect the firing of individual brain cells or neurons, as some in the field have assumed, but the mass electrical activity of groups of cells. That crucial difference has implications for drawing sound conclusions from fMRI data, a difference that leads the author to give the following advice at the close of the introduction:

> Functional MRI is an excellent tool for formulating intelligent, data-based hypotheses [on brain function], but only in certain special cases can it be really useful for unambiguously selecting one of them, or for explaining the detailed neural mechanisms underlying the studied cognitive capacities. In the vast majority of cases, it is the combination of fMRI with other techniques and the parallel use of animal models that will be the most effective strategy for understanding brain function.

MAIN BODY

The main body of the article has two components. The first is a synopsis of fMRI technology and its past uses in brain research; the second concerns what is known about the relationship between neuronal activity detected by fMRI and other analytical techniques and such behavioral responses as arousal and memory retrieval. Throughout the author does not just summarize past literature but makes critical and evaluative observations similar to what one expects in a discussion section (chapter 4). For example,

> functional MRI adaptation designs have been widely used in cognitive neuroscience, but they also have shortcomings, as any area receiving input from another region may reveal adaptation effects that actually occurred in that other region, even if the receiving area itself has no neuronal specific-

ity for the adapted property.[13] Moreover, the conclusions of experiments relying on adaptation designs strongly rely on existing electrophysiological evidence, which itself may hold true for one area and not for another.[72]

At the center of this review is a figure connecting brain function and its fMRI analysis (reproduced here as fig. 7). The circuitlike diagram on the left is a simplified model of the brain illustrating the electrical connections among three populations of cells (boxes inside the graph distributed among six cortical layers listed by roman numerals). The model is dynamic: its operation is initiated by means of an input signal from the thalamus (box outside graph, lower left). The connecting lines in the diagram indicate excitatory synapses (gray) and inhibitory synapses (black); their widths indicate the differing strength of these signals. The bar graphs on the right side of the figure illustrate the categories of fMRI response resulting from "large sustained input changes" (*E* signifies excitation; *I*, inhibition). The first two graphs record responses, respectively, during high and low cortical activity; the second two graphs relate the effects of net excitation and inhibition, measured against a baseline.

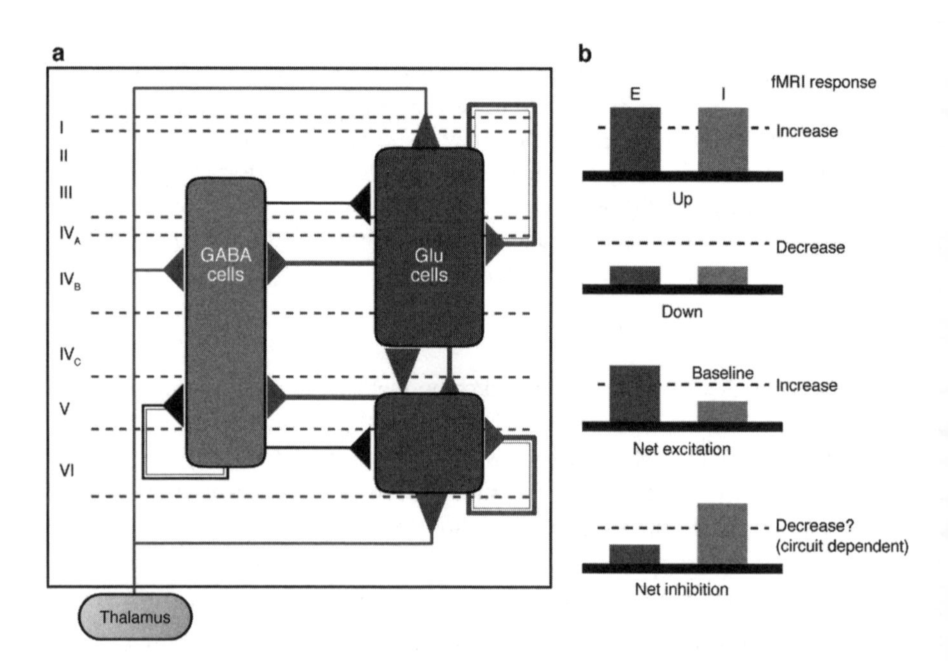

FIGURE 7. Diagram from review article: (a) model of brain activity; (b) typical fMRI responses. Note that our reproduction is gray scale while original version is in color. Reprinted with permission of Macmillan Publisher Ltd. (Logothetis 2008).

At least since the time of Copernicus and his famous diagram of the sun-centered universe, scientists have relied upon visual displays using simplified diagrams as a means of explaining the complex workings of nature. Logothetis's simplified version of the brain on a microscopic scale makes it much easier for his readership to grasp the model whose complex workings his text will examine in detail. Central to his argument is a comparison of his visual model of the cerebral microcircuit with results from published fMRI studies and studies employing analytical methods other than fMRI.

CONCLUSION AND REFERENCES

While the introduction gives the reader an as-yet-unsubstantiated account of what researchers can and cannot do with fMRI, the conclusion presents us with an expanded version of that same statement, fortified this time by the considerable weight of the author's critical analysis of the published literature. As a consequence of this literature review, the author's intuition concerning fMRI's limitations has been transformed into knowledge upon which we can rely. The bases of his conclusions are the 72 articles listed in his reference section, supplemented by the 112 articles listed in a lengthy supplement linked as hypertext to the original article and housed in *Nature* as "supplementary information."

Hybrid Article

We now analyze the workings of a classic 1961 article by François Jacob and Jacques Monod, "Genetic Regulatory Mechanisms in the Synthesis of Proteins." This article is a hybrid: part literature review, part theory. On the one hand, it reviews what was known at the time about the genetic mechanisms controlling protein synthesis within bacteria. On the other hand, it proposes a new theoretical model for the generation of protein from DNA. To encapsulate their theory, the authors coined the term "messenger RNA," a chemical intermediate that carries the genetic information from DNA to protein.

STRUCTURE

This review article is almost forty pages long. For any article of this length, one not governed by a standard format, the key to communicative success is a sound, easy-to-follow structure. In this exemplary and important case, Jacob and Monod organize their content around the processes implicated in their solution to a then-current problem: the lack of a plausible mechanism for protein synthesis. For Jacob and Monod, as their abstract makes clear, protein synthesis involves three types of genes: regulator, operator, and structural.

Structural genes determine the amino acid sequences in the protein. Through the synthesis of enzyme repressors, regulator genes control the transfer of information from structural genes to protein, while operator genes control the formation of the short-lived messenger RNA. Jacob and Monod's main headings, which we have numbered here for clarity, make this argument organizationally visible:

1. Introduction
2. Inductible and Repressible Enzyme Systems
    - The Phenomenon of Enzyme Induction
    - The Lacose System of *Escherichia coli*
    - Enzyme Induction and Protein Synthesis
    - Kinetics of Induction
    - Specificity of Induction
    - Enzyme Repression
    - Kinetics and Specificity of Repression
3. Regulatory Genes
    - Phenotypes and Genotypes in the Lacose Systems
    - The i^+ Gene and Its Cytoplasmic Product
    - Regulator Genes in Repressible Systems
    - The Interaction of Repressors, Inducers and Co-repressors
    - Regulator Genes and Immunity in Template Phase Systems
4. The Operator and the Operon
    - The Operator as Site of Action of the Repressor
    - Constitutive Operator Mutations
    - The Operon
5. The Kinetics of Expression of Structural Genes, and the Nature of the Structural Message
    - Kinetics of Expression of the Galactosidase Structural Gene
    - Structural Effects of Base Analogs
    - Messenger RNA
6. Conclusion
7. References

Notice that Jacob and Monod do not organize their article around one-name-fits-all categories like "methods" or "results," a common practice for experimental articles. Instead, they employ custom-made subheads like "Regulatory Genes" and "The Operator and the Operon." Most long articles do likewise. By examining the headings, readers should be able to perceive the logical arrangement of elements at a glance without having to read the main body. If the headings do not exemplify coherence, the article will likely confuse.

TITLE AND ABSTRACT

The title of this long and complex article is fairly simple and to the point: "Genetic Regulatory Mechanisms in the Synthesis of Proteins." The nucleus noun places the stress on "mechanisms"—the authors are proposing new mechanisms that regulate genes. What abstract would be appropriate for this hybrid of review and theory? If Jacob and Monod had elected to write an abstract that merely summarized their article's contents, a common strategy for long review articles, it might have read as follows:

> The literature has been reviewed on the mechanisms by which genes synthesize proteins in bacteria [from section 1 above and title]. Topics covered include the processes of induction and repression in enzyme synthesis [section 2], the role of regulator genes in the molecular organization of proteins [section 3], and the function of operator genes in information transfer [section 4]. A model is then proposed to explain how all these elements work together to initiate and control protein synthesis [section 5].

Since the authors wanted to emphasize their new theory over their review of the literature, they rejected this format. In their abstract, which we reproduced in chapter 3, they present in a nutshell the new biological mechanism they are proposing, nothing more. Their abstract puts the article's new knowledge claim front and center.

INTRODUCTION

Jacob and Monod's introduction embodies the three steps outlined in chapter 1. In their first sentence they define *gene* as a "DNA molecule whose specific self-replicating structure can, *through mechanisms unknown,* become translated into the specific structure of a polypeptide chain." In the phrase we have just italicized, a gap in current knowledge is implied, one the authors intend to fill. They then go on to state the reigning hypothesis about protein synthesis from past research, one that in their view lacks "experimental support." They formulate two specific problems based on this judgment of evidential inadequacy: (1) Does the synthesis involve only one type of gene or several types, each with a different function? (2) Does the transfer of structural information from DNA to protein involve a "chemical intermediate synthesized by the genes"? The authors close their introduction by summarizing their answers to these questions.

MAIN BODY

Jacob and Monod begin the main body of their article by discussing what is known about the two fundamental processes of genetic control: enzyme repression and induction, the one being the reverse of the other. Following that, separate sections treat the different types of genes and their mechanisms: regulator (section 3), operator (section 4), and structural (section 5). Each section follows the same basic pattern. Within each, the authors put forth numerous hypotheses regarding genetic mechanisms, discuss each in terms of the relevant literature, and reach conclusions concerning which of their hypotheses appear most plausible. They also define new terms, which reflect key concepts of their model for the mechanism of protein formation.

CONCLUSION

The conclusion gives us the final formulation of Jacob and Monod's model. It also differentiates what has been experimentally established from what is speculative and attempts to generalize from the specific bacteria dealt with to higher organisms. Additionally, it hints at future research. It closes with the following eloquent paragraph, yet another summary of the authors' main claims:

> According to the strictly structural concept, the genome is considered as a mosaic of independent molecular blue-prints for the building of individual cellular constituents. In the execution of these plans, however, co-ordination is evidently of absolute survival value. The discovery of regulatory and operator genes, and of repressive regulation of the activity of structural genes, reveals that the genome contains not only a series of blue-prints, but a co-ordinated program of protein synthesis and the means of controlling its execution.

REFERENCES

The last section is a reference list containing 108 sources, 36 of them for research in which either Jacob or Monod or both participated. In other words, much of this article explains facts acquired in their previous experimental work.

The Digital Article: Wave of the Future?

The standard arrangement we defined in the previous chapter evolved in the course of the twentieth century. But evolution does not end just because a century ends. The scientific article continues to evolve, and the main driving force of change at the moment is new technology in the form of the personal computer and Internet.

Digital research articles on the Web sites of current journals are composed of two communicative layers. The first and most prominent consists of an article little different from its printed journal form; the second layer, accessed by means of hyperlinks, is a trove of supplementary information for any specialist readers seeking more detail. It also may contain a communication designed for general readers. We will illustrate both layers with an example from *Nature* (Green et al. 2006).

VISIBLE LAYER

In "Analysis of One Million Base Pairs of Neanderthal DNA," Richard E. Green and ten others (2006) tell how they sequenced a large chunk of the genome of the extinct hominid group known as Neanderthal. On the Web or in print you will find an article that looks and reads much like any other scientific article. It has an abstract (indicated by bold type, no heading), an introduction (no heading but implied because it immediately follows the abstract), a methods section, results and discussion sections, a conclusion section, references, and acknowledgments.

The abstract conforms to convention; it establishes why this research is important to the advance of evolutionary biology, what the authors discovered, and how they discovered it. The introduction focuses on what was known about Neanderthals through archaeological and DNA studies and establishes the authors' fundamental research problem: the need for better understanding of genetic differences between humans and Neanderthals, differences that will provide insights into how humans evolved. The next section establishes a methodological problem: how to identify a Neanderthal fossil that has not been contaminated by exposure to human handling. Having solved that problem, the authors turn to a second: how to determine the DNA sequence in an uncontaminated thirty-eight-thousand-year-old Neanderthal fossil. With those methodological problems resolved, the authors interpret their DNA sequencing results by comparison with the genomes of modern humans and other primates. Their key discovery is that the DNA sequences for Neanderthals diverged from that of humans about half a million years ago. Their conclusion elaborates on the ways in which this research can be extended.

HIDDEN LAYER

Examine the Web version of the article closely, and you will find some intriguing differences:

- At the top of the Web version is a table of contents linked to the appropriate sections of the article.
- There are links to earlier *Nature* articles on the same topic, as well as to the references cited in the text. Click on a reference number in the main

body, and you will advance to the relevant bibliographic information from the print version. Click at the end of the bibliographic information, and up pops the cited article.

- Appended to the bibliographic information are hyperlinks that track citation records, made available through the ISI Web of Knowledge and ChemPort.

At the viewers' fingertips, there is, courtesy of the Internet, a small library of the relevant published literature.

At the end of the article is the tag "Supplementary Information." Click on that and you will find additional methods, results, tables of data, and figures. In fact, the main article itself reads somewhat differently from its nondigital ancestor in that the bulk of the methods description and acquired data appears in this supplementary material—not the sections on methods or results in the main body. Because of this electronic arrangement, journals can publish a streamlined version of an article meant to minimize the reading time for an audience mainly interested in what was discovered and what arguments support it, reserving the methods details and much of the data to hyperlinks designed for the few specialists who might be building upon the discovery or questioning its validity. With these special-interest details hidden away, the result is a somewhat more reader-friendly article than would likely have resulted by the conventional route.

The supplementary material in the selected article ends with a hyperlink to a streaming video aimed at a general audience. In this *Nova*-like video, we learn how three of the ten scientists (Ed Green, Johannes Krause, and Adrian Briggs) came to work on the Neanderthal genome, why their discovery is important to our understanding of human evolution, how they made their discovery, exactly what they discovered, and what scientific breakthroughs we can expect as a result. In other words, the streaming video covers much the same ground as the article itself but with a much different audience in mind. Having viewed the video on the Web, general readers are much better armed to read the accompanying scientific article, if they so choose.

We foresee a day when such articles are the norm:

- They benefit all readers, who first confront a relatively short article with fewer technical details than the present norm.
- They benefit experts, who have easy access to full details on methods, results, and other supplementary information that might assist them in judging the current article and directing their own research.
- They benefit authors, who have the opportunity to reach more readers by writing a less technical version of their article, while at the same time

making supplementary material easily available for specialists and nonspecialists alike.

In this brave new digital world, scientist-authors will be called upon to more frequently adjust their text and visual displays to reach different audiences. Our chapters 11, 12, and 13 offer advice on how to do just that.

Conclusion

When taking a course in college physics, students typically prepare for exams by learning to solve problems chosen because they clearly illuminate the principles being studied. Exams then test the students' ability to apply these examples in a creative way to similar but different problems. Good exam questions do not permit students simply to choose the right formula from the textbook, plug in the data given in the questions, and grind out the answers. They require them to think creatively. Analogously, you should think of writing a scientific article as a test in which you are called upon to solve a set of expository problems by creatively applying the principles in part I. You pass with flying colors if your first choice of journal or publisher accepts your paper, either as it is or subject to revision.

The purpose of this chapter is to emphasize the important point that the scientific research article exists in forms other than the one represented in the previous chapter. The scientific article is not a species but a genus, not a genre but a family of genres. Each member of this family exists and flourishes because it is well adapted to its purpose; its form is as much an exercise in intelligence as is its content. And as the rise of the digital article illustrates, the scientific article continues to evolve, to adapt to changing circumstances.

PART II

Beyond the Scientific Article

10 Proposing New Research

Scientific articles and grant proposals share many similarities. Problems are defined. Methods are outlined. Experiments are described. Citations are amassed. But there are also differences. The subject of this chapter is these differences. An abstract of a grant proposal by Abraham Loeb (Harvard-Smithsonian Center for Astrophysics) and Rennan Barkana (Tel-Aviv University) underlines the essential distinction. The grant ("Observable Signatures of Reionization") concerns the history of the early universe. Here is its abstract:

> Current observations indicate that the hydrogen in the intergalactic medium was still neutral at redshift [a unit of relative astronomical distance] 30, but was already highly ionized by redshift 6. This implies that the first sources of light, most likely stars and quasars that began to form around redshift 30, produced a phase transition of hydrogen reionization in this redshift interval. We propose to make detailed predictions of signatures of the reionization era, with the goal of clarifying how upcoming observations can be used to study the era of reionization and learn about the process of galaxy formation as well as properties of the dark matter. The proposed work will combine semi-analytic calculations and limited numerical simulations with radiative transfer, focusing on physical situations or reionization scenarios where full cosmological simulations are not feasible. The predictions will be tested over the next decade with observations from space-based missions, including *SIRTF* and *NGST*, *MAP*, *PLANK*, *Chandra*, and *XMM*, ground-based optical/infrared telescopes such as *Keck*, *VLT*, and other 10 meter or larger telescopes planned for the coming decade, and planned radio telescopes such as the Square Kilometer Array.

As the words we have underscored make clear, the shift from scientific article to proposal is a shift from the past to the future, from work accomplished to accomplishment proposed, from a judgment of the quality of the work to a judgment of the credibility of the researcher and the proposed research. The reason for this shift is obvious: the higher the credibility of the researchers and their proposed research, the more likely they will be successful in achieving its goals. For the researcher, however, this shift creates a dilemma: because credibility is established on the basis of past performance, it is a conservative force, emphasizing development of an established line of inquiry over innovation, Thomas Kuhn's normal over his revolutionary science. This means that scientists' most speculative ideas may have to survive initially without funds or be smuggled into existing grants.

Worries about this dilemma were a persistent topic of conversation between us and the scientists who graciously shared their work and thoughts with us. For example, Terry Hunt of the University of Hawaii said concerning his fascinating hypothesis on the decline of Easter Island civilization: "I have feared, and I may have been wrong all the time, that research which challenges the status quo is difficult to fund. So I have not put my crazy ideas out there in grant proposals. We (Carl Lippo & I) feel that given our recent paper in *Science* our colleagues may now have to listen to us, and a grant proposal may now have a [good] chance in peer review."

We will leave to philosophers, historians, sociologists, and proposal review panels the question of what exactly constitutes a healthy balance between "sound" and "wild" ideas in the funding of research. In this chapter, we will focus instead on the ways established scientists communicate their credibility and the credibility of the research they propose, the ways they set at ease the minds of members of review panels and funding agencies.

In this chapter, we will ignore the formal characteristics of the grants (organization, font size, length) on the grounds that each funding agency has its own guidelines, guidelines that are readily available and that it is clearly wise to follow as closely as possible. Why lessen your chances of success just because you did not choose the right font size or exceeded the maximum length? However, no matter what the stated guidelines, we can generalize that all proposal readers expect answers to three fundamental questions:

1. What is the researchers' credibility, given the proposed problem?
2. Is the researchers' method for solving the proposed problem credible?
3. What will be the scientific and social benefits of having solved the problem?

The Credibility of the Researchers

The credibility of the researchers has three bases: the curriculum vitae of each member of the research team, the track record of relevant preliminary research from the researchers' laboratory, and the degree of potential intellectual synergy resulting from a history of collaboration, sharing, and interaction among grants already bestowed.

The credibility of a scientist's curriculum vitae rests on two evidential pillars: relevant biographical facts and relevant publications. Adrian Bejan, for example, introduces himself by listing his degrees and his professional positions, ending with his current one, J. A. Jones Distinguished Professor of Mechanical Engineering at Duke University. He then lists his areas of expertise. At this point he reaches the autobiographical heart of his resume, the professional achievements that form the basis for the trust he hopes that the review panel will bestow:

> Professor Bejan is the author of 20 books and 440 journal articles. He is the recipient of the Max Jakob Memorial Award of the American Society of Mechanical Engineers (ASME) and the American Institute of Chemical Engineers. He received the Ralph Coates Roe Award of the American Society of Engineering Education. From ASME he also received the Worcester Reed Warner Medal, The Edward F. Ober Award, the James Harry Potter Gold Medal, the Charles Russ Richards Memorial Award, the Gustus L. Larson Memorial Award, and the Heat Transfer Memorial Award in Science. He was awarded the Luikov Medal by the International Center of Heat and Mass Transfer.
>
> Adrian Bejan was awarded 15 honorary doctorates at universities in 10 countries, e.g., the Swiss Federal Institute of Technology Zürich 2003.
>
> He is ranked among the 100 most cited authors in all of engineering, all fields, all countries.

In these paragraphs Bejan lists number of publications, awards, honorary degrees, and professional ranking based on numbers of citations received. Taken together, they make a strong case that this particular researcher has reached the highest level of achievement in his field.

Relevant publications are the second evidential pillar of the curriculum vitae's credibility. In his proposal for the renewal of his grant to investigate the causes of Alzheimer's disease, Michael Wolfe of Harvard University lists the following publications, all of which flowed from the initial grant:

1. M. S. Wolfe, M. Citron, T. S. Diehl, W. Xia, I. O. Donkor, and D. J. Selkoe. "A substrate-based difluoroketone selectively inhibits Alzheimer's γ-secretase activity." *J. Med. Chem.*, 41, 6–9 (1998).
2. M. S. Wolfe, W. Xia, C. L. Moore, D. D. Leatherwood, B. Ostaszewski, I. O. Donkor, and D. J. Selkoe. "Difluoroketone peptidomimetic probes for Alzheimer's γ-secretases: Evidence for loose sequence specificity and an aspartyl protease mechanism." *Biochemistry*, 38, 4720–7 (1999).
3. *M. S. Wolfe, W. Xia, B. L. Ostaszewski, T. S. Diehl, J. Shen, and D. J. Selkoe. "Two conserved transmembrane aspartates in presenilin 1 are

required for both presenilin endoproteolysis and γ-secretase activity." *Nature,* 398, 513–7 (1999).

4. *B. De Strooper, W. Annaert, P. Cupers, P. Saftig, K. Craesserts, J. S. Mumm, E. H. Schroeter, V. Schrijvers, M. S. Wolfe, W. J. Ray, A. Goate, and R. Kopan. "A presenilin-1-dependent γ-secretase-like protease mediates release of notch intracellular domain." *Nature,* 398, 518–22 (1999).

This is no ordinary list. All three journals are among the most frequently cited in his discipline. Moreover, the asterisks beside the third and fourth articles are not mere window decoration: these are "among the ten most cited papers of 1999."

Research credibility is also enhanced by collaboration and sharing among laboratories and by interaction among grants themselves. Here a military analogy may be appropriate. Collaboration and sharing are "force multiplying factors"; they assure funding agencies that their resources are being maximized. Indications of collaboration are a constant in the grant proposals we studied: Shane Ross of Virginia Tech speaks of "a recent collaboration of myself with chemists and other mathematicians"; Terry Collins of Carnegie Mellon mentions working on spectroscopic analysis with the assistance of a colleague; Kip Hodges and Kelin Whipple of Arizona State devote a special section to "Relationship to Other Research Initiatives": "A variety of researchers from other institutions are carrying out studies in the Arequipa regions. . . . Our work dovetails nicely with theirs inasmuch as it is critical for our evaluation of the aggradational terrace results that we understand the role of landsliding in producing the alluvial fill in the upper Colca valley." Sharing between laboratories also multiplies the force of research. Michael Wolfe offers to make the pharmacological agents he develops in his Alzheimer's research available "freely . . . to other academic researchers." Interaction among the existing grants is another force multiplier. Allan I. Basbaum of the University of California, San Francisco, refers in his proposal to an initiative in his laboratory, "an incredibly fruitful neuroanatomical investigation of neurotransmitter receptors." While this investigation is relevant to the work for which he has requested funding, it is the product of another of his grants, already funded.

In short, key to any grant proposal is making the case that your research team has the capacity to carry out the proposed research, a successful project that will reflect well on the funding agency. To that end, we recommend that you do not mindlessly paste in your standard curriculum vitae and list of publications but arrange and edit them to accentuate the match between your credentials and the research being proposed.

The Credibility of the Research

The credibility of the research depends on accomplishing four tasks, three of which are shared by scientific articles as well as grant proposals. First, researchers must define a research problem and depict it as a step forward in the discipline. Second, they must suggest a proposed solution to the problem. Third, they must detail the social and scientific impacts of the solution they propose. But while the purpose of these tasks in a scientific article is to justify the correctness and significance of the authors' solution, their purpose in a grant proposal is to make credible its possibility, in effect to turn its possibility into a probability. The fourth task is unique to grant proposals: presenting a realistic time line and a justified budget.

THE PROBLEM AS A STEP FORWARD

Successful scientific proposals impress review panels by proposing an important step forward, one that advances a research front significantly. They propose to accomplish this advance by filling a need, closing a gap, or resolving a problem, a contradiction, or an inconsistency. To achieve this end, a question must be located precisely in a research front. Here is how Shane Ross's grant proposal "Multiscale Dynamics and Phase Transport in Nonintegrable Dynamic Systems" accomplishes this task by a thorough survey of the latest research:

> Dynamical systems theory has experienced considerable growth towards applications, mainly motivated by recent progress in the development of numerical techniques for dynamical problems and the availability of more powerful computational facilities. However, challenging problems remain. One area of particular interest is *phase space transport* [author's emphasis], a unified mathematical description of dynamical processes which can be applied to a wide range of physical phenomena across many scales, such as atomic physics [26], physical chemistry [4, 5, 41, 64], fluid mixing [22–24, 34, 47], climate models [1, 45, 46], low energy spacecraft trajectory design [20, 21, 29, 30, 56, 57], asteroid and comet evolution [14, 29, 31, 33, 53–55, 58], stellar motion [18, 25], and cosmological models of large scale mass distribution in the universe [6, 7].
>
> These physical systems can be modeled initially as non-integrable dynamical systems, and many are Hamiltonian with $n \geq 2$ degrees of freedom (n dof). Simulation and theoretical understanding of this rich class of problems is important to many areas of science and engineering. To realistically model some problems requires $n \geq 3$ dof [33, 65], i.e., a phase space of six dimensions or more. This high dimensionality has made the systematic study of such systems difficult.

In this brief space, Ross accomplishes three important goals: he gives us the history of his research front, integrates his own published work into that front (references 14, 20, 21, 29, 30, 31, 33, and 53 58), and states the problem he proposes to tackle. In stating the problem, he moves from the general to the specific, a very common tactic. Ross starts with a fast-moving research front (dynamical systems theory), moves on to a narrower area within that front (phase space transport), and concludes with his particular research problem—the extension of dynamical systems theory to more realistic multidimensional phase space.

THE PROPOSED SOLUTION

Successful research proposals also impress review panels by setting forth a plausible approach to solving the problem. Ross characterizes his problem as "difficult." How is this difficulty to be addressed? This brings us to his proposed fourfold solution to the problem: (1) merging the theories of tube and lobe dynamics, (2) establishing a firm theoretical link between the geometry and statistics, (3) working within the context of two example systems, one from molecular, the other from fluid dynamics, and (4) accurately calculating phase space transport quantities for the two example systems. In our discussion, we will focus on the first research objective:

> MERGE THE THEORIES OF TUBE AND LOBE DYNAMICS [29–33, 49–52] into a single geometric theory for multiscale transport in Hamiltonian systems. This has largely been completed for the 2 dof case [54]. I will focus my efforts on n 3 dof, where no such comprehensive theory has yet been formulated.

Ross's exposition of his method in his figure 1 (see our figure 8) increases credibility by demonstrating a mastery of a mathematical technique, using tube and lobe dynamics to derive a sample trajectory. But it also does this by employing a set of diagrams from a published article of which he is the author. Publication represents a community judgment that the work on which he is basing his proposal has met disciplinary standards. (In this figure, "RTBP" refers to the restricted three-body problem, restricted because a general solution to its dynamics has never been found.)

THE IMPACT OF THE SOLUTION

Proposal writers enhance the credibility of their proposed research by detailing its potential impact on society or science or both. While in introductions to scientific articles this step is sometimes left out or barely mentioned, it takes on greatly increased importance in proposals. Evaluators of proposals must be convinced that solving the limited research problem under consid-

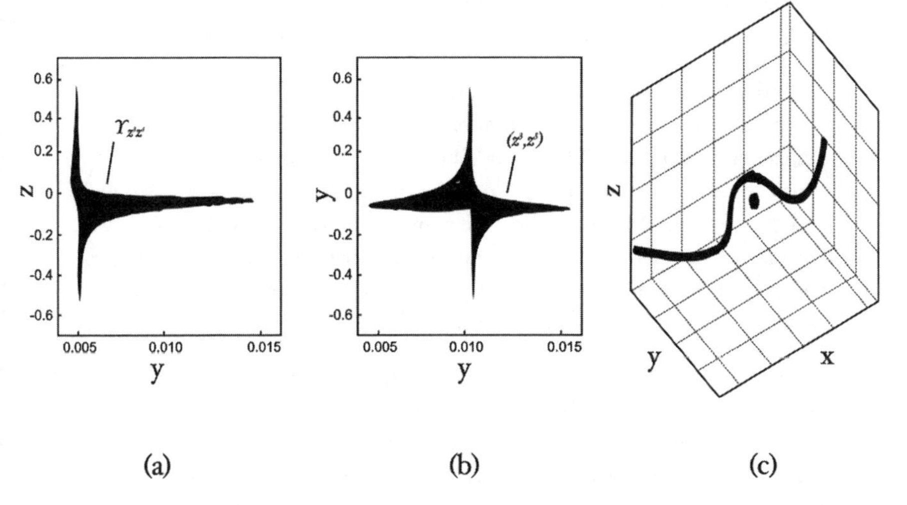

On a 5-dimensional energy surface, there are 4-dimensional invariant manifold "tubes" (local structure, $S^3 \times R$). Taking a 4-dimensional Poincaré surface of section which intersects a tube, one finds a 3-dimensional separatrix with a S^3 structure. Two 2-dimensional projections of the object, (a) and (b), are shown. To find points inside the separatrix, one does the following: Due to the S^3 structure, every point $(z',\dot{z}')$ in (b) corresponds to a closed curve $\gamma_{z',\dot{z}'}$ in (a), as in the example shown. Fixing the point in (b), one picks a point inside the region bounded by $\gamma_{z',\dot{z}'}$ in (a). This numerical example is from the 3 dof RTBP (reproduced from [21]). (c) The configuration space projection of an example trajectory in the RTBP which goes between three realms.

FIGURE 8. Graph from research proposal: Tube dynamics in a system with three degrees of freedom (dof). Reprinted with permission of author (Ross 2003).

eration will reflect favorably upon the funding agency they represent. You should not assume that review panels will immediately recognize what the ultimate benefits are. At the same time, you must take care not to overstate your research's importance, as such assertions can weaken your credibility. Review panels may well doubt that your research will single-handedly solve the world's energy problems, explain the origin of the cosmos, or cure cancer.

Michael Wolfe's approach to this problem is exemplary. His work on the role of γ-secretase in Alzheimer's disease (AD), he contends, will lead to new treatments for this increasing problem in our rapidly aging society. It will do so by leading the pharmaceutical industry away from a costly and frustrating trial-and-error approach: "As γ-secretase is an important target for the potential treatment of AD, understanding the topography of the active site and identifying residues that compose it should facilitate rational drug design." In addition, his research will advance the science of biochemistry:

> Identifying specific residues in the active site of γ-secretase should also help answer a more fundamental biochemical question. γ-Secretase cuts in the middle of the APP transmembrane domain, and the two aspartates critical for activity are thought to lie in the middle of presenilin TM6 and TM7[41,62]. . . . However, direct demonstration that the active site . . . is intramembranous has not been reported. If our transition-state analogues bind to residues that are clearly in presenlin regions, this would be strong evidence for an intramembranous active site.

Note that Wolfe does not claim broadly that his work will lead one day to a cure for Alzheimer's; he elaborates on exactly on what his specific research might contribute toward taming that devastating disease.

You can also add credibility to your proposal by citing its potential contribution to the skills of students and the knowledge of the general public. Academic review panels, in particular, like to learn about how the proposed research will contribute to the training of the next generation of scientists. A proposal by Thomas D. Seeley of Cornell University on decision-making in bees makes the importance of student contributions particularly clear. In his original proposal, he states that he "involves students in the field data collection stage of research." In the revision requested by the review panel, he is more specific: "I should have made it clear that nearly all the co-authors on papers for which I am the senior author are undergraduate students who helped me perform the research being reported. In the past four years, these assistants have been Susannah C. Buhrman, Alexander S. Mikheyev, Gary J. Pagano, and Anja Weidenmüller. I involve them in the data analysis and the writing of the papers; hence I include them as authors, as you will see from the list of publications below." Seeley makes it clear that, simultaneously, he is doing research and training potential researchers.

Similarly, Kip Hodges and Kelin Whipple increase the credibility of their grant proposal on the uplift history of the Peruvian Cordillera Occidental mountain range by suggesting that the general public would benefit from their research:

> We are very excited about a planned informal educational initiative that would be done in collaboration with Marc Goddard of Bio Bio Expeditions. In the course of discussions prior to the submission of this proposal, we developed an idea to coauthor a popular science book about the geomorphology of rapids, in effect an effort to explain river hydrodynamics to recreational rafters. While we would use the Colca and Cotahuasi as ex-

amples of river phenomena, the book would be aimed at a broad spectrum of readers. Hopefully, we will be able to encourage in the general public a greater appreciation of the role of tectonics in shaping the evolution of river systems.

TIME LINES AND BUDGETING

Realistic research plans and time lines add to credibility by emphasizing the researcher's good judgment. They permit review panels to gauge whether your ambitions meet or exceed what you can reasonably accomplish. You start with a research plan: a set of objectives, tasks, or aims you expect to complete within a time frame, usually three to five years. Allan Basbaum's plan is an example. His research tests various hypotheses concerning pain pathways; to do so, he alters the genes of mice. In the time line below, these are designated by various letter-number combinations (PPT-A and NK-1, SP and NKA nulls, αCAMKIIT286A). This time line testifies to Basbaum's ability to judge with relative precision the temporal durations of "specific aims I through IV" over the proposed grant period of five years:

Year 1	Year 2	Year 3	Year 4	Year 5

SPECIFIC AIM I

Electrophysiology of PPT-A and NK-1 mice

SPECIFIC AIM II

Create constructs

Target deletion

Generate mice with single deletion

Behavioral analysis of SP and NKA nulls

Neurochemistry of SP and NKA nulls

Electrophysiology of SP and NKA nulls

Fos induction / NK-1 receptor studies

SPECIFIC AIM III

Injury induced changes in second messengers

Screen for changes by immunocytochemistry and *in situ*

Comparison of changes in PPT-A, NK-1R null and wild type mice

SPECIFIC AIM IV

Behavioral analysis of αCaMKII null and αCaMKIIT286A mice

Electrophysiology of αCaMKII mice

Realistic budgets also add to a researcher's credibility. Budgeting does not consist merely of aligning cost items to dollar amounts. Budgets must often be justified in the text itself. In Terry Collins's "green chemistry" proposal, for example, he not only lists the staff involved but also explains their roles, as follows:

1. The senior investigator "oversees the development of all scientific projects and collaborations as well as the day-to-day management activities."
2. A research associate "leads the day-to-day in laboratory studies involving mechanistic investigations into . . . the underlying chemical reasons for the superior performance of TAML activators [catalysts that facilitate the neutralization of pollutants] in a variety of environmentally important processes such as the bleaching of dyes, wood pulp, and colored effluents, the degradation of persistent organic pollutants, and the decontamination of chemical and biological warfare agents."
3. Two graduate students "will participate in the on-going mechanistic research and carry out specific sub-system studies in described fields-of-use."

A travel budget for these participants is justified by indicating that these trips are designed to permit presentation of the results of the proposed grant at scientific conferences.

Accountability applies also to equipment and supplies. It is important that these budget items are specific to the grant, that the grant is *not* for purchasing items for general laboratory use. Accordingly, Collins assures the granting agency that "these costs can be identified with a high degree of accuracy" and that they include only such "technical supplies . . . as solvents, gases, lab supplies, chemicals, instrument charges, and equipment maintenance necessary to perform the described research."

Dealing with Resubmission

Not all grants are funded in the first round of reviews. In fact, most first proposals on a new line of research get rejected on the first try. In fortunate cases, the review panel will request a resubmission in response to criticisms.

When criticized, many of us immediately become defensive—accusing our critics of willful misreading, ignorance, nit-picking, pigheadedness, and even worse. Let that mood pass, then reexamine the criticisms again in a more sedate mood. You will likely find some merit in them. Few if any writers are so skilled that the documents they produce are beyond criticism.

Fred L. Gould is a highly respected researcher at North Carolina State.

Nevertheless, his proposal concerning insect pest control through mutation was initially rejected. One of the review panel's concerns was the skills available in Gould's laboratory. Gould notes in his successful resubmission that although his laboratory "has experience with the use of genetic markers for [the] purpose of [creating] native transgenetic mosquito strains," he is well aware of his group's limitations. Accordingly, he mentions that a member of his research network, Thomas W. Scott of the University of California, Davis, is collaborating with Dana A. Focks, a retired government entomologist now in the private sector, "on a related project, so interaction will be facilitated." Gould is also aware that the grid models that his laboratory has favored have limited applicability. To address this limitation, he proposes to extend his research network to include Alun Lloyd, a colleague in the Biomath program: "Lloyd brings expertise in network models to our project. He will work on alternative model frameworks including network models, and we will conduct head on comparisons of the efficiency and outputs of the grid and alternative models when using the same biological parameter values."

The crux of the review panel critique, however, concerns the credibility of Gould's population genetics model. His response is not to jettison the model but to incorporate tests of its credibility into the grant proposal itself. Accordingly, his resubmission features a recurrent heading, "Model Credibility":

> MODEL CREDIBILITY: Sensitivity to environmental conditions. By conducting a sensitivity analysis for each of the transgenic strategies described above we will be able to determine if the success of some strategies is expected to be more affected by varied biological and physical conditions than others. Sensitivity analysis will also enable us to determine if some strategies are always better than others. If the models indicate that variation in conditions do not affect the ranking of strategies, this indicates that the model predictions are likely to be very robust. If the best approach is highly dependent upon environmental conditions then empirical work will be needed to carefully assess local conditions before any release, and whether modifications of the genetic approaches are needed.

Open disagreement with a review panel is, obviously, infrequent. In a proposal concerning the advance of "green" chemistry, however, Terry Collins tactfully dissents from the extended time line proposed by the panel: "It has been mentioned in the Panel Summary that the time frame of the project could be extended to 1.5–2 years. After the submission of the first version of the project, the research at the Institute for Green Oxidation Chemistry did not stop at all and significant new results were obtained. . . . We believe that one year will give sufficient time to accomplish the major goals. The newly added list of tasks will show that our goals are in fact realistic for one year's funding." In his dissent, Collins ingeniously employs as an argument for retaining his original time line the time-lag incident on panel review. Devising credible responses to criticisms, as Collins and the others quoted in this section did, is part of establishing your credibility.

Other Types of Grants

Grant proposals to further research are not the only type; funds can also be requested for pedagogical projects. For example, in his "New 4th Year Undergraduate Course on 'Constructal Design of Energy-System Configuration,'" Andrian Bejan proposes to convey to undergraduates a theory he has developed on energy transport. He argues for his pedagogical project as follows: "[T]he objective of the proposed course is to use the existing research base, and to bring the latest design ideas into the classroom. I believe that this is the best way to prepare engineers: teach fundamentals and disciplines first, and emphasize freedom, novelty, and the interdisciplinary next. In constructal design, interdisciplinary considerations recommend themselves 'naturally' when designers encounter systems with more than one objective." At this point, Bejan outlines a method for achieving his results:

1. Anchor the new material in appropriately selected classical material that has features in common with the new material,
2. Use the simplest possible formulations and applications of the new material,
3. Use creative graphics, and teach one more time (through examples) elements of the basic disciplines and math,
4. Encourage the student to think freely and to *contribute ideas* to the development of the new material and eventual textbook [his emphasis].

In the course, these principles would be applied to a wide variety of examples, from lung design and river basin morphology to refrigerator design and traffic flow. Finally, course development would not cease at the grant's termination; Bejan would continue to develop the course through the use of various forms of student feedback.

For this project, Bejan presents himself as a credible principal investigator:

> Constructal design is my most recent and, I would say, most promising method that deserves to become part of the engineering curriculum. I have

> developed at least two earlier research methods into teaching methods that have become accepted worldwide: thermodynamic optimization, or entropy generation minimalization, and scale analysis of convective heat and mass transfer. Throughout my career I made a strong effort to bring the latest research ideas and techniques into the classroom.

Funding can also be requested for individuals. David Jewitt and Shane Ross have both sought funding for this purpose—in Jewitt's case for his graduate student, in Ross's case for himself. In both cases, the proposers spend considerable time demonstrating the value of their programs of research. In this respect, their grants resemble those we have already explored in detail; indeed, we have already used Ross's for this purpose. The *specific* credibility of these proposals, however, concerns potential career impact. Jewitt says that participation will provide his graduate student "with a good grounding in planetary astronomy. As is my usual style, I will work with the student closely on every part of the research, from taking the data, reducing it, understanding its meaning and publishing the results. Although often very exhausting to me, this is by far the best way to make sure that a person will learn how to do science well and is, in the long term, undoubtedly worth the effort."

In his postdoc proposal, Ross emphasizes its potential impact on his professional life: "The work I propose here," he says, "will give me a chance to build a solid resume and reputation at this critical early point in my career. My goal is to work in academia, performing basic research in dynamical systems and control, while teaching a new generation of scientists and engineers the tools and insight that come with viewing dynamics geometrically." To further this goal, Ross proposes research that "lays a good foundation for learning and developing such techniques in the context of two important problems in molecular dynamics and fluid dynamics, i.e., the TAM [tri-atomic molecule] and TVP [three vortex problem]. Furthermore, understanding of these two 'simple' problems lays the groundwork for systematically building and understanding more complicated models." His choice of postdoc mentor, he feels, furthers his goal substantially: "Working with Paul Newton of the University of Southern California (USC), particularly on the N-vortex problem and geophysical fluid dynamics, will allow me to gain the experience and contacts I need to develop the fluids side of my work. Furthermore, making links with experts in other fields at USC will broaden my perspective on how the methods I am developing could be applied to important and computationally intensive problems."

Concluding Remark

The grant process is extremely competitive; many proposals fail. In the best of all possible worlds, resources would be allocated so as to maximize the overall quality of the science that would be produced. As a consequence, in a merit-based system we might expect that the best scientists with the best research records will always win out over their rivals. So construed, this system seems to place those starting out at a considerable disadvantage. Everyone realizes, however, that, if those starting out are not given the resources they need to progress, there will be no next generation of senior research scientists; everyone recognizes that well-trained young men and women must be supported to maintain disciplinary health.

EXERCISE

Like the research article, most full-length proposals open with an abstract. Key to success is creating an abstract that not only summarizes the proposed research but touches upon the credibility of the researchers and the research they propose. Below we have created a very short abstract for an imaginary grant proposal based upon James Watson's autobiographical account *The Double Helix,* by far the best personal account of the circumstances leading up to a major scientific discovery:

> Deoxyribonucleic acid is a molecule whose structure has so far eluded solution, despite considerable efforts over a long period of time [PROBLEM]. The researchers—a biologist and a physicist—intend that the synergy of their combined training succeed where others have failed [COLLABORATION]. To reconstruct a three-dimensional structure within the steric constraints that the x-ray diffraction photographs reveal, they will employ principles from mathematical physics, such as Fourier transforms [SOLUTION]. At the same time, their hope is that the structure of this molecule once elucidated will shed light on its biological significance, especially its relationship to problems of heredity [IMPACT].

Now it is your turn to create an abstract for a test case. Imagine that you have been working on a new ceramic material for the geological disposal of hazardous waste. Your preliminary tests have shown that this material is so strong, durable, and easy to make that it could replace conventional concrete for many other purposes like building construction and road repair. Moreover, a quick calculation suggests that worldwide replacement of concrete with this new ceramic could reduce greenhouse gases by as much as 5 percent because no carbon dioxide is released to the environment when it is made. You still need to run tests to determine the optimum ceramic composition and perform more rigorous economic calculations. With those facts at your disposal, can you write a persuasive abstract of proposed research?

CHECKLIST

Whether you are a junior or senior research scientist, whether your research proposal is a few pages or hundreds, whether your budget is in the thousands or millions of dollars, you will need to persuade a review panel of experts that you have identified a research problem worth solving, have devised a credible plan for solving it, and possess the capabilities and resources to execute the plan on time and within budget. To do so, we suggest asking the following questions of your draft proposal or resubmission:

Researcher Credibility

- Is your curriculum vitae tailored specifically to the research proposal you are submitting for funding?
- Is the work of the laboratory with which you are associated presented in a light that maximizes its relevance to your proposal?
- Have you maximized collaborations with other laboratories that might be relevant to the completion of the research proposed?

Research Credibility

- Have you clearly stated the problem you intend to solve?
- Have you clearly outlined why you believe the methods by which you intend to solve the problem will work?
- Have you made clear the ultimate impact of your proposed research on your discipline, society, the training of your students, and, if appropriate, the knowledge of the general public?
- Have you proposed a reasonable time line and a fully justified budget?
- In resubmissions, have you responded clearly and positively to the criticisms of the review panel? When forced to disagree with the panel, have you done so tactfully?

For other types of grants in the sciences, our suggested questions differ:

- In the case of pedagogical grants, do you have a rationale for the course of instruction, an outline of the curriculum, a specification of the methods you intend to employ to implement that curriculum, and a means of evaluating its success?

- Have you presented yourself as an obvious choice to develop such a course?
- In the case of grants for individual support, havc you made it clear that your career plans coincide with the advance of the research front with which you have allied yourself?
- Have you presented the laboratory and its lead scientist in a way that makes clear the positive impact both will have on your future as a researcher?

11 Going Public

A classic magazine cartoon features a wealthy patron visiting a sculptor's studio and asking whether his work is difficult. "Not at all," he replies; "I simply purchase a block of marble and chip away the parts I don't need." For the rules we are about to give you, this caveat applies. The rules are simple, but following them represents a serious challenge.

These rules concern how scientists can turn their research into a sequence of words and pictures that is attractive and comprehensible to interested nonscientists, science administrators, science teachers, and scientists who are not specialists in that particular subdiscipline. For most scientists, this writing task is only an occasional requirement of their job. And in our experience, scientists tend to find this task bothersome and even question its value. Why should they make a special effort to reach a wider audience?

We see three advantages. First, writing for the public identifies the authors as spokespersons for their discipline. Second, writing for a public that includes scientists with much different training can further interdisciplinary endeavors: we need to remember that the discovery of the structure of DNA was the work of both a physicist and a biologist, neither of whom had by himself the answer to the puzzle. Finally, and perhaps most important, when well done, such writing leaves a favorable impression on the public—in particular, the citizens whose taxes support much of scientific research. Writing for the general public celebrates science.

This chapter offers six "rules," or, to be more exact, suggestions, designed to increase the reader-friendliness of scientific communication aimed at general audiences. We derived these suggestions from a close reading of a sampling from *Scientific American* and *American Scientist.* In our view, the scientist-authors and their expert editors in both journals routinely avoid the errors that bedevil bad science writing for the general public. In general, these articles make for engaging reading yet do not simplify or strip the science bare to the point of distortion. At the same time, they do not make unreasonable demands regarding the readers' familiarity with the subject at hand.

Developing a Good Story

Suggestion 1: Follow a simple overall plan, a structure consisting of three elements: context, problem, solution.

The typical specialized article follows the overall plan detailed in chapter 8: a title and abstract that capture the key points of the whole article, an introduction that establishes a limited research problem within a wider research territory, the method applied to solving that problem, results from having applied the method, some discussion of the significance of those results, a conclusion that reiterates key points and touches upon possible directions for continued research, acknowledgments that list sources of financial and intellectual support, and a reference list. The typical popular science article follows a similar though simpler plan—one we call "context-problem-solution." We can convey a sense of how this plan works by quoting from the overview to "The Strangest Satellites in the Solar System" by David Jewitt, Scott Sheppard, and Jan Kleyna (2006). This overview is meant to stand on its own, independent of the rest of the article (below and elsewhere in this chapter, the italicized headings are our addition):

Context

- Astronomers used to think that most planetary moons formed from disks around their respective planets—reproducing, in miniature, the formation of the solar system itself. These moons orbit in the same plane as the planet's equator and in the same direction as the planet's spin. The few bodies not fitting this pattern were deemed "irregular." A recent flood of discoveries using advanced digital detectors shows that irregular moons are actually the majority. Their long, looping, slanted orbits indicate that they did not form in situ but instead in paths encircling the sun. In essence, they are asteroids or comets that the planets somehow captured.

Problem

- Neither the source region nor the mechanism of capture is well understood.

Provisional solution

- The moons might have come from the Kuiper belt beyond Neptune or from regions closer in. Their capture may have involved collisions or other interactions in a younger, more densely populated solar system.

In the full article, an engaging title, a headline, and a series of headings reinforce the context-problem-solution framework while at the same time they grab the reader's attention:

Title
The Strangest Satellites in the Solar System

Headline
Found in stretched, slanted loop-d-loop orbits, an odd breed of planetary satellites opens a window into the formation of planets.

Context and problem
Black Sheep
Cosmic Polyrhythm

Solutions
What a Drag
Three's a Crowd
Planetary Movements

Note that the headings exploit catchphrases, or apt variations on them, as a means of drawing the reader into the text. Still, the journalistic quality of the headings makes it difficult to see their close relationship to the article's structure. To do so, we translate them into what linguist Michael Halliday calls "scientific English":

Context-problem
Irregular orbits of many planetary satellites

Solutions
Explaining These Orbits: The Gas-Drag Hypothesis
Explaining These Orbits: The "Pull-Down" Hypothesis
Explaining These Orbits: The Three-Body-Capture Hypothesis

Which style do you prefer? In our view, as long as the first few sentences following the headings clarify their meaning, attention-grabbing titles and headings are worth concocting. Nonetheless, while journalistic titles and headings are popular, they are not required. In an article in *American Scientist* on the spread of malaria and other mosquito-borne diseases, Fred Gould, Krisztian Magori,and Yunxin Huang (2006) use somewhat more technical language than do Jewitt, Sheppard, and Kleyna:

Title
Genetic Strategies for Controlling Mosquito-Borne Diseases

Headline
Engineered genes that block transmission of malaria and dengue can hitch a ride on selfish DNA and spread into wild populations

Untitled introduction covering context and problem
Solutions
Transgenes and Fitness
Strain Replacement
A Two-Transgene Technique
The Uses of Selfish DNA
Simple Eradication
Social Context and Risk

Getting Started

Suggestion 2: Give your article an attractive title and headline that convey the main message in plain language.

Nowadays the titles of specialized scientific articles are very specific—so much so that they form a sort of miniabstract, a summary of what is to follow (see chapter 3). In marked contrast, the titles of science articles for nonspecialist audiences tend to be short and inventive, often drawing on common phrases from popular culture or employing such literary devices as puns, metaphors and similes, personification, and alliteration. The main job of these titles is not to inform but to entice. Once enticed, however, readers need immediately to be informed. Both *Scientific American* and *American Scientist* follow such titles with a short sentence meant to capture the article's main message in a nutshell, what we call a "headline."

Thomas Quinn, Andrew Hendry, and Gregory Buck (2001) titled their specialized-audience article on bear predation "Balancing Natural and Sexual Selection in Sockeye Salmon: Interactions between Body Size, Reproductive Opportunity and Vulnerability to Predation by Bears." An article in *Scientific American* based on the same research was alliteratively titled "The Fish and the Forest" (Gende and Quinn 2006). To this catchy but vague title is added a headline specifying the article's content: "Salmon-catching bears fertilize forests with the partially eaten carcasses of their favorite food." Charles Conley titled his specialized article on space travel "Low Energy Transit Orbits in the Restricted Three-Body Problem" (1968). An *American Scientist* article based on the same research was titled "The Interplanetary Transport Network" (Ross 2006). To this metaphorical and evocative title is added a headline that gets right to the article's main message: "Some mathematical sophistication allows spacecraft to be maneuvered over large distances using little or no fuel."

In both examples, the title and headline work together to capture readers' attention without sacrificing content. Readers of the specialized scientific literature read because staying up to date is part of their job. By and large, they scavenge the literature to learn new techniques, experimental results, and theoretical explanations that will help them find and solve new research problems. Readers of scientific articles for nonspecialists have no such strong motivation. For the most part, they read because a title or its headline has piqued their curiosity. They must be enticed to read with the promise that they might learn something new and interesting.

How do you create a catchy title if you are not practiced in doing so? For inspiration, look to the titles of highly successful popularizations by first-rank scientists: Richard Feynman resorted to a pun in *QED*, Brian Greene personified the cosmos in his *Elegant Universe*, Stephen Jay Gould borrowed phrases from popular culture in *Wonderful Life* and *Full House*, and in *The First Three Minutes* Steven Weinberg alluded to Genesis.

Suggestion 3: Begin your article with a fact, situation, or anecdote designed to build a bond between you and your potential readers, then introduce the problem or discovery.

According to linguist John Swales (see chapter 1), modern introductions in specialized publications conventionally do so in three stages:

1. [*Define research territory*] This stage normally summarizes the state of knowledge in the scientific research front being studied.
2. [*Establish problem*] In this stage, authors point out a contradiction or inconsistency or gap in that state, or propose to build upon a neglected, undeveloped, or misunderstood aspect of it.
3. [*Possible solution*] This final stage summarizes or suggests a solution to the problem.

To discover the differences between such introductions and those of science articles for non-specialists, we will compare the introductions of two articles with parallel subject matter: Craig Agnor and Douglas Hamilton's "Neptune's Capture of Its Moon Triton in a Binary-Planet Gravitational Encounter" (2006), appearing in *Nature;* and Jewitt, Sheppard, and Kleyna's "The Strangest Satellites in the Solar System" (2006), appearing in *Scientific American.* The first article's introduction easily exemplifies Swales's three stages:

> *Research territory*
> Triton is Neptune's principal satellite and is by far the largest retrograde satellite in the Solar System (its mass is ~40 per cent greater than that of Pluto). Its inclined and circular orbit lies between a group of small inner prograde satellites and a number of exterior irregular satellites with both prograde and retrograde orbits.

Problem
This unusual configuration has led to the belief that Triton originally orbited the Sun before being captured in orbit around Neptune.[1,2,3] Existing models[4,5,6] for its capture, however, all have significant bottlenecks that make their effectiveness doubtful.

Solution
Here we report that a three-body gravitational encounter between a binary system (of ~10^3-kilometer-sized bodies) and Neptune is a far more likely explanation of Triton's capture. Our model predicts that Triton was once a member of a binary with a range of plausible characteristics, including ones similar to the Pluto-Charon pair.

The introduction to the typical article aimed at a general audience also consists of three stages:

1. [*Hook*] This stage provides a tantalizing detail to link the article's content to its audience. This may be a personal anecdote or an issue within the general reader's experience. The purpose of the hook is to dramatize the problem being addressed and, simultaneously, to link the reader's experience and interests to the article's subject matter.
2. [*Background*] This stage summarizes the relevant work in the research field.
3. [*New research*] This stage indicates new research that bears on a problem with which the article will deal.

You may have noticed some similarities between these stages and those of the introduction in the standard scientific article; actually, the last two are versions of the first three, adapted to a general audience. Only the hook is really new. We recommend its use for a general audience, since authors cannot assume that their audience is already motivated.

Here are selections from Jewitt, Sheppard, and Kleyna's introductory material in "The Strangest Satellites in the Solar System" (2006):

Hook
Five years ago two of us whiled away a cloudy night on the summit of Mauna Kea in Hawaii by placing bets on how many moons remained to be discovered in the solar system. Jewitt wagered $100 that a dedicated telescopic search could find, at most, 10 new ones. After all, he reasoned, in the entire 20th century, astronomers had come across only a few. Sheppard more optimistically predicted twice as many, given the increased sensitivity of modern astronomical facilities.

Sheppard is now a richer man . . .

> *Background*
> Since that night, our team has discovered 62 moons around the giant planets, with more in the pipeline. Other groups have found an additional 24. . . . No one predicted that the family of the sun had so many members lurking in the shadows. They are classified as "irregular," meaning that their orbits are large, highly elliptical and tilted with respect to the equators of their host planets. So-called regular moons, such as Earth's or the large Galilean satellites of Jupiter, have comparatively tight, circular and nearly equatorial orbits.
>
> *New research*
> These bodies are not well explained by standard models, and a wave of fresh theoretical work is under way. It seems that they are products of a long-gone epoch when the gravitational tug of the newly formed planets scattered—or snatched—small bodies from their original orbits.

All three stages need not appear in an introduction, nor need they conform to the order given. You must tailor them to what works best for the particular case. For example, Thomas Seeley, Kirk Visscher, and Kevin Passino (2006) combine the hook and problem statement in an *American Scientist* article on the social behavior of bees:

> The problem of social choice has challenged social philosophers and political scientists for centuries. The fundamental decision-making dilemma for groups is how to turn individual preferences for different outcomes into a single choice for the group as a whole.

Finishing with a Flourish

Suggestion 4: In your conclusion, do not merely sum up; also talk about your science's future and its wider implications.

In a manner parallel to Swales's three-step introduction, we break down the conclusion of the standard scientific article into three stages (see chapter 5):

1. [*Reiteration*] In this stage, authors support original claims with the evidence in the previous text.
2. [*Wider significance*] In this stage, authors mention the wider significance of those claims to the research territory under scrutiny.
3. [*Future work*] In this stage, authors touch upon possible future work that would validate or make use of the original claims.

To exemplify the conclusion to a standard scientific article, we turn to the article in *Evolutionary Ecology Research* by Quinn, Hendry, and Buck (2001) on bear predation of salmon, a source paper for the general-audience article "The Fish and the Forest" (Gende and Quinn 2006):

> *Reiteration*
> In conclusion, we note the prevalence of premature mortality in these creeks (largely from bears) and in many other streams in Alaska and British Columbia . . .
>
> *Wider significance*
> [We] firmly believe that bears subject many Pacific salmon populations to substantial predation. Such predation may have been an important factor shaping the life history, morphology, breeding phenology and behaviour of salmon, and . . .
>
> *Future work*
> . . . [C]ontrolled studies should explicitly consider the effects that predation would have had on the results.

The full conclusions of science articles aimed at nonspecialists are analogous. They differ in that they add a stage—an optional segment that points to a moral. Conclusions of this sort give more emphasis to a project's relevance to larger societal concerns, an emphasis seldom found in the specialized literature.

1. [*Reiteration*] In this stage, authors sum up the original claims in the previous text.
2. [*Wider significance*] This stage scrutinizes the wider significance of those claims to science or society or both.
3. [*Future work*] In this stage, authors touch upon possible future work that would validate or make use of the original claims.
4. [*Moral*] This stage may include practical uses, lessons learned, or ethical implications of the solution. It may also recommend social policies.

While the first three concluding stages are basically the same for the specialized and general article, the moral differs in that it can readily drift beyond the purely scientific. This conclusion, from a *Scientific American* article by Allan Basbaum and David Julius, "Toward Better Pain Control" (2006), gives us the first three stages, while, at the same time, it places their research in a cultural context wider than that in which standard scientific articles are set:

> *Reiteration*
> In this article we have discussed a subset of the experimental approaches to treating pain, all of which have shown promise in animal studies. Those evoking the greatest excitement leave normal sensation intact while diminishing

the heightened sensitization characteristic of the difficult-to-treat inflammatory and neuropathic pains and have an acceptable side-effects profile . . .

Wider significance
But will these therapies help patients? And will they work on all types of pain?

Future work
One approach that deserves further exploration is the use of behavioral, non-drug therapies for intractable pain—particularly that associated with conditions such as fibromyalgia and irritable bowel syndrome, for which no one has conclusively established an organic cause . . .

Moral
Poet Emily Dickinson often contemplated pain. In one work, she noted

> Pain has an element of blank;
> It cannot recollect
> When it began, or if there were
> A day when it is not.
> It has no future but itself.

We can only hope continued research into the mechanisms of pain sensation will lead to safe, effective treatments that will alter pain's future, such that it reverts to a time when it was not.

Good conclusions like the above try to leave readers feeling they learned something new and of potential importance not just to a small group of researchers working in a highly specialized area but to a much wider community.

Also note the cautious language of the final quoted sentence: "We can only hope . . . " That final paragraph ensures that general readers will not draw any unwarranted conclusions about the utility of the research. In the attempt to attract a wide audience by emphasizing the positive, scientist-authors (and science journalists for that matter) must always guard against misrepresenting the degree of certainty behind the research or the time it will likely take for some invention to make a difference to the readers' lives.

Developing a Public Writing Style

Suggestion 5: Adjust your writing style by defining central technical terms, incorporating informal language into formal prose, and employing figures of speech like metaphor not only to enliven and explain but also as organizing principles.

The prose in specialized journals is technical, formal, and impersonal. Good science writing for general audiences regularly departs from those trends. As a result, journals like *Scientific American* and *American Scientist* exhibit a stylistic freedom that is absent from standard scientific prose. Here is an example on the origin of solar flares by Gordon Holman (2006):

> The weather on earth, complicated as it is, at least results from familiar processes: solar heating, differences in air pressure and shifting wind patterns. So most people have an intuitive grasp of why, for instance, the skies can be sunny one day and rainy the next. In contrast, solar flares and other aspects of "space weather" involve the interplay of magnetic fields and gas that is hot enough to become ionized (which is to say that the constituent atoms are stripped of their electrons). Such interactions cannot be seen directly and can be tricky to visualize, even for specialists. The leading idea for how these goings-on generate solar flares—magnetic reconnection—dates back to the 1950s and 1960s. Yet observational evidence for it has been slow in coming, so much so that some space physicists were beginning to have their doubts about the theory's merit.
>
> Scientists generally agree that the energy released in a flare must first be stored in the sun's magnetic fields. That surmise follows from the fact that the flares erupt from parts of the sun called active regions, where solar magnetic fields are much stronger than average. These areas are most easily identified by the presence of sunspots—those dark-looking patches host the most intense magnetic fields on the sun. In these zones, the lines of force of the magnetic field extend from the surface into the corona, the outer layer of the solar atmosphere, arching upward in broad loops, which trap hot gas—and I do mean *hot:* several million kelvins.

Holman is not talking down to his readers. The style of this passage is predominantly though not exclusively formal and impersonal: the subjects of his sentences are almost invariably solar flares and related phenomena. The passage is also full of technical terms: "ionized," "solar flares," "magnetic fields," "magnetic reconnection," "sunspots," "corona," "kelvins." Concerning the meaning of these terms, Holman assumes that the reader possesses a certain degree of scientific literacy: for example, he does not define "magnet fields" or "kelvins." But he clearly presumes a nonspecialized audience: "ionized," "sunspots," and "corona" are defined, as is "solar flares" (in a previous passage). The definition of "magnetic reconnection"—the article's central term—will unfold as the article unfolds.

Despite this passage's formality, the personal can erupt at any time: "and I do mean *hot*" is an example. In addition, the passage is interspersed with expressions characteristic of informal prose: "at least" in the first sentence; "most people," "sunny," and "rainy" in the second; "space weather" and "hot enough" in the third; "tricky" in the fourth; "goings-on" and "dates back" in the fifth; "slow in coming" in the sixth.

In the prose of science articles for nonspecialists, there is also a tendency to convey meaning through such figures of speech as personification, metaphor, and simile. In "Group Decision Making in Honey Bee Swarms," Seeley, Visscher, and Passino (2006) personify, portraying bees as human actors: "There is no doubt then that the dancing bees were reporting nest sites. Indeed, it seemed these bees were holding a kind of plebiscite on the swarm's future home, although exactly how they conducted their deliberations was still unknown."

Because metaphor and simile work by picturing one thing in terms of another, they are also helpful for conveying difficult ideas by making the unfamiliar familiar. For example, Holman compares "space weather" with "earth weather," bringing the heavens down to earth. In a *Scientific American* article on the use of "green chemistry" to reduce waterway pollution, Terence Collins and Chip Walter (2006) advocate for a future world in which the old slogan "better living through chemistry" rings true: "The advances of *green chemistry* to date represent only *a few interim steps on the road* to dealing with the many environmental challenges of the 21st century" (our emphasis).

Metaphor and simile can also become a central principle in an article's organization. In writing for popular consumption, historian of science Peter Galison (2006) recommends that science writers "find a felicitous metaphor and stick with it—using the conceit to display the interconnectedness of the phenomena that at first glance appear scattered within a scientific domain." In harmony with Galison's advice, Shane Ross (2006) uses hydrodynamic metaphors to explain a new concept in space travel. We are told "the solar system turns out to be more like a turbulent sea than a clockwork." Nevertheless, we can journey through space by doing "what sailors have long done—taking advantage of ocean currents to speed them where they want to go." We can take advantage of the complex geometry of space and the effects of gravity: our space vehicle begins "to trail L_2 [a Lagrange point], and encounters another rising surface behind, which acts to speed [the space vehicle] up and to scoot it toward the Sun, just as a wave propels a surfer toward the beach (and often a little sideways)." Finally, we are told that "the comparison with fluids is more than just an analogy." This use of metaphor corresponds exactly to Galison's recommendation. It brings together otherwise apparently disparate features into a single expository structure.

Nor need this use of metaphor be confined to physics. The already cited *Scientific American* article by Scott Gende and Thomas Quinn, "The Fish and the Forest" (2006), concerns the ecology of salmon harvesting. Its thesis is that bears and other predators and scavengers return to the soil nutrients that are essential to the preservation of ecological balance. The metaphor compares the nutrient cycle to traffic, an organizational metaphor mirrored exactly in the article's headings: "The Nutrient Express," "Special Delivery," and "Managing the Nutrient Express."

Making Science Visible to the Public

Suggestion 6: Adapt your tables and illustrations to a general audience by means of strategic simplifications and amplifications.

In scientific articles proper, authors closely coordinate tables and illustrations with text. They number every illustration and table, and each is referred to by a number at the appropriate juncture in the text. Readers are not expected to read the text alongside its illustrations and tables but to move back and forth between these two very different representations of scientific knowledge, each of which conveys essential information that does not overlap completely with the other. In scientific articles aimed primarily at general audiences, the situation is different. Illustrations and tables simply reinforce the message of the text.

In "The Fish and the Forest," for example, text, table, and illustrations convey the message that bear-salmon predation is vital to the preservation of Alaskan forest ecology, a balance that sometimes must be fostered by artificial means. The table and illustrations make this process vivid. We see what the scientists saw: the bear and the salmon, the spread of nutrients to other forest animals from the salmon carcasses resulting from bear predation, the bear-to-bear rivalry that sometimes inhibits this predation, and a helicopter delivery of salmon carcasses to the Baker River in Washington State. We also see some of the data on which the scientists erected their hypotheses: a bar chart of the nutrients contained in salmon and a table on the extent of bear predation.

How do tables in general-audience science articles differ in structure from their standard counterparts? Figure 9 is a table for a specialized audience from Quinn, Hendry, and Buck 2001. This table contains seven categories of information: year, sex, creek of origin, number of salmon, and cause of death by median and percentage. It has twenty rows of data, nine columns, and one footnote. Within the main body of the text, this table is incorporated into a complex statistical argument concerning the extent of bear predation of salmon and the role of natural and sexual selection.

In stark contrast is the far simpler *Scientific American* table (Gende and Quinn 2006): see figure 10. This table contains only four categories of information: name of creek, average number of salmon in it, average num-

Year	Sex		Cause of death					
			Bear-killed	Missing	Stranded	Gull-pecked	Premature	Senescent
Hansen Creek								
1999	Female	% ($n = 38$)	7.9	39.5	2.6	2.6	52.6	47.4
		median	3	2	1	1	2	12
	Male	% ($n = 93$)	9.7	29.0	5.4	3.2	47.3	52.7
		median	2	2	1	1	2	13
2000	Female	% ($n = 57$)	66.7	17.5	5.3	0.0	89.5	10.5
		median	2	2	2	—	2	3.5
	Male	% ($n = 117$)	54.7	26.5	11.1	0.0	92.3	7.7
		median	3	2	1	—	2	5
	Total	% ($n = 305$)	37.4	27.2	7.2	1.3	73.1	26.9
		median	2	2	1	1	2	11
Pick Creek								
1995	Female	% ($n = 114$)	5.3	34.2	0.0	0.0	39.5	60.5
		median	10	17	—	—	14	22
	Male	% ($n = 119$)	21.0	68.1	0.0	0.0	89.1	10.9
		median	13	8	—	—	9	17
1996	Female	% ($n = 167$)	1.8	18.6	0.0	7.2	27.5	72.5
		median	6	12	—	9.5	10	18
	Male	% ($n = 156$)	6.4	45.5	0.0	0.0	51.9	48.1
		median	14.5	12	—	—	13	18
	Total	% ($n = 556$)	7.9	39.9	0.0	2.2	50.0	50.0
		median	13	11	—	9.5	11.5	19

Note: The category 'Premature' combines bear-killed, missing, stranded and gull-pecked fish.

Percentage of tagged female and male sockeye salmon in two creeks, sampled in two different years, that were killed by bears, stranded, pecked by gulls, missing, or dying a natural (senescent) death, and the median number of days in the stream of each of these groups.

FIGURE 9. Table meant for scientists: Data on salmon mortality in two Alaskan creeks. Reprinted with permission of author (Quinn, Hendry, and Buck 2001).

ber killed by bears, and average percent killed. It has only nine rows and four columns. The reader need really concentrate only on the first and last columns to get the main point, where the percent killed ranges from 12 to 58; the main point is that bears kill a very high percentage of the salmon in most Alaskan creeks. Its legend summarizes a more general point: the salmon "population densities are so great that the fish have a huge impact on freshwater systems." Popular science tables like this one work best when they make one main point with minimal scanning, matching, integrating, or interpreting. (For a more detailed discussion, see chapter 12.)

BEAR PREDATION

Creek	Average no. of salmon	Average no. killed by bears	Average percent killed
Bear	3,907	1,183	32
Big Whitefish	786	342	48
Eagle	818	399	53
Fenno	5,228	666	12
Hansen	6,229	2,450	49
Hidden Lake	2,010	671	43
Little Whitefish	173	93	58
Pick	5,837	1,949	35

FIGURE 10. Table meant for the general public: Data on bear predation of salmon in Alaskan creeks. Designed by Lucy Reading-Ikkanda for *Scientific American Magazine* (Gende and Quinn 2006).

How do illustrations in popular science articles differ from their standard counterparts? In the *Bulletin of the American Mathematical Society* (Marsden and Ross 2005), we have a qualitative representation of a mathematical solution to the orbital problem faced by the spacecraft (fig. 11). Here is how the authors work this scientific visual into the argument of the article:

> When we consider a spacecraft with control instead of a comet, we can intelligently exploit the transfer dynamics to construct low energy trajectories with prescribed behaviors, such as transfers between adjacent moons in the Jovian and Saturnian systems. . . . In an earlier study of a transfer from Ganymede to Europa, we found our fuel consumption for impulsive burns to be less than half the Hohman transfer value. We found this to be the case for the following example of a multi-moon orbiter tour that is shown schematically in Figure [10.1]: starting beyond Ganymede's orbit, the spacecraft is ballistically captured by Ganymede, orbits it once, escapes in the direction of Europa, and ends in a ballistic capture at Europa.

In Shane Ross's "Interplanetary Transport Network" appearing in *American Scientist* (2006), a variation of this same diagram appears (see fig. 12). At first glance, this illustration (provided in color in *American Scientist*) seems much more complex than its original: in addition to the satellite orbits, it includes representations of the gravitational deformations of space figured as "tubes," and a detail that focuses on the orbit around Europa controlled by the Lagrange point, L_2, marked by an X. But it is complexity in the service of ease of understanding. To make their point, the authors use color and contrast to represent clearly the different trajectories of the spacecraft and Jupiter moons. They add an enlarged version of the complicated path the spacecraft takes to its final destination. And they supplement this vivid visualization with explanatory text interpreting the figure. The whole is more complex in construction but simpler in comprehension for a nonexpert; it is, in fact, a lucid explanation and visualization of a complex process. In the transition, however, the illustration has lost both its mathematical anchor and its argumentative function. Its role is now solely expository.

We consider three steps to be essential in making the transition to tables and illustrations in popular science articles. First, strip away any technical details of sole interest to specialists. You must be ruthless here. If you are not sure, it can probably go. Second, draw upon all textual and visual means available to explain the contents, especially color, legends, and labels. Third, make sure that everything in the table or illustration is self-explanatory or explained in plain words.

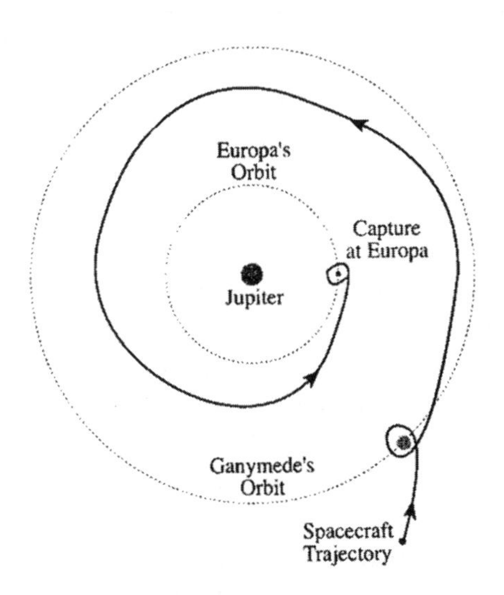

FIGURE 11. Figure meant for scientists: Leap-frogging mission concept of multi-moon orbiter tour of Jupiter's moons Ganymede and Europa. Reprinted with permission of American Mathematical Society (Marsden and Ross 2006).

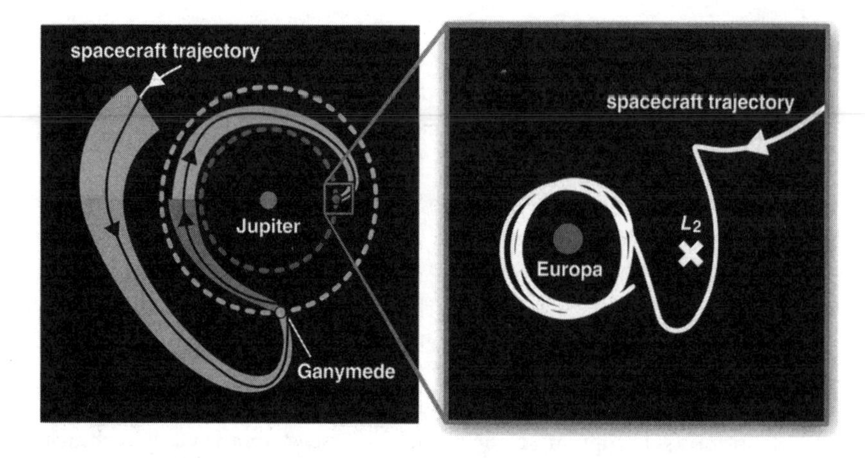

Exploration of Jupiter's icy moons could benefit from a cleverly designed trajectory. A probe could, for example, enter the Jovian system along an inbound tube (*outer swath beginning at "spacecraft trajectory"*) that carried it toward Jupiter's moon Ganymede, which it would orbit briefly before following an outbound tube (*expanding inner swath beginning at Ganymede*) that conveyed it into an orbit around Jupiter that was smaller than Ganymede's. The probe would then hop to an inbound tube toward Jupiter's moon Europa (*contracting inner swath ending at Europa*), which it would then orbit for a significant time (*right*).

FIGURE 12. Figure meant for the general public. Note that our reproduction is gray scale while original version is in color. Designed by Jen Christiansen (Ross 2006).

Conclusion

We recommend that all researchers have on their computer's hard drive a short, up-to-date article (say, 2,000–3,000 words) about their research program, aimed at a general audience. To get in the mood to compose this article, imagine yourself at a dinner party. Someone asks, "What do you do?" and "Does that actually matter in the real world?" We view having a general-audience version of your research on hand as being at least as important as having an up-to-date curriculum vitae for research proposals. It can be drawn upon and modified whenever you are called upon to concoct a "reader-friendly" version of your research. It can also be called upon in a pinch at a dinner party.

EXERCISE

The top box on page 152 reproduces the abstract from a scientific article on evolutionary biology. At the bottom is our attempt to translate it into prose understandable by a bright high school student surfing the Net. It opens with a catchy title, headline, and hook, just as we have recommended. It closes with a moral to the story. Try a similar exercise with a scientific article of your choosing—preferable one on a topic in which you have a special interest. Search the Web for any helpful background information.

CHECKLIST

In the process of translating complex science into understandable prose and pictures for a general audience, try to do the following:

- Concoct an attention-getting title and an informative headline.
- Organize your thoughts around a context-problem-solution structure that tells a compelling story of discovery. If at all possible, support that organization with a central metaphor that will aid the understanding of lay audiences.
- Begin with a "hook"—a fact, situation, or anecdote designed to build a bond between you and your potential readers.
- End with talk about your science's future, its wider significance, and policy or ethical implications.
- For applied research, address at some point how close it is to practical use by others. For controversial research, discuss any new knowledge claims that others might question.
- Adopt a writing style that is less formal and more personal than that of scientific articles proper, being careful to avoid mathematics, if at all possible, and to define any difficult technical terms.
- Finally, make your story vivid by means of a judicious selection of tables and illustrations. Adapt any originally intended for a specialized audience to its general counterpart by means of strategic simplifications and amplifications. Add legends that clearly explain the purpose of these tables and illustrations.

Abstract: "Balancing natural and sexual selection in sockeye salmon: interactions between body size, reproductive opportunity and vulnerability to predation by bears."

Traits that increase reproductive success, such as body size and sexual dimorphism, may compromise survival, leading to opposing pressures of natural and sexual selection. Discrete populations exposed to different balances between selective forces should differ in phenotypic traits associated with natural and sexual selection. We used two proximate populations of sockeye salmon (*Oncorhynchus nerka*) that differ in body size as a model for studying this kind of balancing selection. We hypothesized that large body size would enhance potential reproductive success through relationships with duration of nest guarding in females, and dominance and duration of reproductive life in males, but that it would be opposed by probability of premature death, chiefly from predation by bears. Longevity on the breeding grounds was primarily controlled by predation, which varied between creeks and years. Pick Creek salmon experienced less predation than those in Hansen Creek and also tended to live longer before being killed, giving Pick Creek females a higher probability of completing egg deposition and males a greater opportunity to breed than those in Hansen Creek. In addition, Hansen Creek salmon were subjected to strong, size-selective predation and also selective mortality from stranding as they ascended the mouth of the creek, whereas we found no evidence of size-selective mortality among Pick Creek salmon. Male dominance in courtship for access to females favoured large salmon, except when predation was very intense. These patterns of balancing selection were consistent with the larger body size of sockeye salmon in Pick Creek. We also found that premature mortality, especially predation by bears, can significantly truncate the reproductive opportunities of salmon, raising a cautionary note regarding controlled studies in which predation cannot occur. (Quinn, Hendry, and Buck 2001)

Bigger Is Not Always Better
When bears eat salmon, they add to our knowledge of natural selection.

On the elementary school playground, bigger is always better: the bully always wins. Not so in the wild—or at least not necessarily. Ordinarily, you'd think that the bigger the male sockeye salmon, the more likely that he will mate, and the bigger the female salmon, the more likely it would be that she will guard the nest successfully and so be chosen as a mate. In other words, sexual selection would operate unopposed by any counterforce. But in the wild there is such a counterforce: predatory bears. Look at the fate of sockeye salmon in two contrasting sites in Alaska: Pick Creek, where predation was generally light, and Hansen Creek, where it was generally heavy. In Hansen Creek, there was evidence that bear predation led to the survival of salmon of smaller sizes—bears naturally prefer to catch larger salmon—while in Pick Creek no such evidence emerged. What are the conclusions of such a study? First, where bear predation was intense, sexual and natural selection operated at cross-purposes. Second, caution is advisable in taking at their face value those studies of reproductive behavior in salmon that do not take predation into account. Third, one ought not take for granted what at first glance might appear obviously true.

12 Presenting PowerPoint Science

At a minimum, a presentation format should do no harm. Yet the PowerPoint style routinely disrupts, dominates, and trivializes content. Thus PowerPoint presentations too often resemble a school play—very loud, very slow, and very simple.

Edward Tufte (2003)

No handbook on scientific communication would be complete without a discussion of scientific talks, and no discussion of scientific talks would be complete without discussing PowerPoint, a milestone in communication media preceded in importance only by paper, the blackboard, the whiteboard, the overhead, and the slide projector. Unlike these other milestones, however, this one has been accompanied by extensive criticism, most prominently the tsunami of criticism generated by the justly admired Edward Tufte. In *The Cognitive Style of PowerPoint* (Tufte 2003), he has made PowerPoint responsible not only for the *Columbia* disaster but also for far more widespread communicative disasters in business meetings and lecture halls. In Tufte's own words:

> How is it that each elaborate architecture of thought *always* fits exactly on one slide? The rigid slide-by-slide hierarchies, indifferent to content, slice and dice the evidence into arbitrary compartments, producing an anti-narrative with choppy continuity. Medieval in its preoccupation with hierarchical distinctions, the PowerPoint format signals every bullet's status in 4 or 5 different simultaneous ways: by the order in sequence, extent of indent, size of bullet, and size of type associated with various bullets.

A common failing of PowerPoint presentations in science, however, is one that Tufte fails to discuss: their creators fail to adjust the contents to take into account that their audience has only a minute or two to view each slide. Our diagnoses of the source of this problem is that scientists too often paste tables and figures designed for journal publication into PowerPoint slides with little or no modification to adjust for the new medium of expression. In this chapter, we offer advice on how to adjust the level of visual and cognitive processing required for each slide with respect to auditors as distinct from readers. In doing so, we show how to avoid the traps that PowerPoint sets, traps that Tufte so perspicuously analyzes.

Tables

To understand a particular cell or group of cells in a table involves two tasks: scanning and matching. Each scan and match requires visual attention accompanied by analytical thought. Figure 13 is a fairly typical table from a scientific journal article in molecular biochemistry (Taylor 1960). Here is one way to read this complex table:

1. We scan from the table title, "Autoradiographic Data on Incorporation of Cytidine-H^3 [tritium] into RNA of Chinese Hamster Cells in Culture," to the column supertitle, "Total grains with standard errors over nucleus or cytoplasm and an equivalent area without cells (background).*"
2. To understand the column supertitle, we scan to the asterisk footnote and back again.
3. Then we scan over to the column subtitle on the left, "Time after incorporation began," and down to the two categories of strain, A1290 and 1404.
4. At this point, scanning per se ends and matching begins. We choose a strain and within that a particular row, let us say Strain A1290 at "10 + 10 min." (From the text, we understand that the first "10" is the time in minutes the cells were in contact with a medium containing a radioactive marker of tritium, while the second "10" is the time they were in a medium free of that marker.)
5. Then we choose a particular column, let's say the first, and scan up to its identifying label, "Nucleus," then down to the data on grains incorporated, then over again to the time elapsed.

The product of all of this scanning and matching is the sentence "In the case of Strain A1290, at 10 + 10 minutes the total grains in the nucleus equals 77.0 ± 1.8." We then match that number with its adjacent background count: 2.9 ± 0.8 grains. To those in the know, that leads to the sentence "The background-corrected total grains in the nucleus equals 74.1" (ignoring the standard deviation). Tables are a way of arranging verbal and numerical information so that the author can efficiently generate a large number of such parallel sentences.

But tables are also ways of discovering data patterns. A scan down the fourth column, for example, the one labeled "Cytoplasm," shows a pattern of numerical increase that may be interpreted as supporting the sentence "The genetic material RNA is transported to the cytoplasm in a cell over time"—a

TABLE 1
AUTORADIOGRAPHIC DATA ON INCORPORATION OF CYTIDINE-H^3 INTO RNA OF CHINESE HAMSTER CELLS IN CULTURE

Time after incorporation began	Total grains with standard errors over nucleus or cytoplasm and an equivalent area without cells (background)*			
	Nucleus	Background	Cytoplasm	Background
Strain A1290				
5–6 min.	57.0 ± 2.1	2.0 ± 0.4	3.0 ± 0.3	4.0 ± 0.3
10 min.	86.3 ± 2.1	2.9 ± 0.4	3.3 ± 0.3	6.4 ± 0.6
10 + 10 min.	77.0 ± 1.8	2.9 ± 0.8	5.2 ± 0.4	5.0 ± 0.5
10 + 30 min.	57.3 ± 1.7	3.4 ± 0.4	25.2 ± 1.2	7.4 ± 0.7
10 + 60 min.	{56.1 ± 2.1 / 82.5 ± 2.3}†	3.0 ± 0.4	{58.6 ± 3.0 / 67.6 ± 2.3}†	6.6 ± 0.8
10 + 2½ hours	{34.5 ± 2.4 / 69.6 ± 2.1}†	2.4 ± 0.6	{106.1 ± 5.9 / 127.5 ± 6.4}†	6.0 ± 0.5
10 + 4 hours	{25.8 ± 2.5 / 85.8 ± 3.8}†	3.4 ± 0.3	{115.8 ± 8.1 / 139.5 ± 7.4}†	6.9 ± 0.5
Strain 1404				
5 min.	34.9 ± 1.5	1.2 ± 0.3	2.3 ± 0.3	4.3 ± 0.7
10 min.	64.1 ± 2.4	1.8 ± 0.5	4.3 ± 0.4	6.1 ± 0.6
10 + 10 min.	65.0 ± 3.3	0.8 ± 0.3	6.9 ± 0.4	3.4 ± 0.4

* Average number of grains over 50 cells and background count over 10 areas of the same average area as the nucleus or cytoplasm.

† Since the amount of labeled DNA in the nucleus raises the count considerably after one half hour, the cells were divided into two classes: the 56 per cent with the lowest counts and the remaining 44 per cent (number estimated to be in DNA synthesis during the 10-min. contact).

FIGURE 13. Table meant for scientists: Data on incorporation of radioactive marker into RNA of hamster cells. Reprinted with permission of Blackwell Publishing (Taylor 1960).

corollary of the major claim of the article, that RNA is synthesized in the nucleus. Tabular presentation, then, not only facilitates the search for new information and its interpretation but also facilitates its integration into the article as a whole.

What if Herbert Taylor wanted to transform this table, designed for publication in a scientific journal, into a PowerPoint slide? To do so he would want to reduce the mental processing that he could legitimately ask of audiences as distinct from readers. That might involve dropping the standard deviation and background data so that his main point stands out clearly. It might also involve dropping the footnotes, simplifying the headings, and adding a slide title that expresses the main point in a short sentence.

We now have a table (fig. 14) that requires only three columns of data to present the evidence in favor of the point of its title. You may object to the loss of statistical data and subordinate technical details—understandably so. But which table would you want to have to decipher in a few minutes' time? Some compromises must be made to avoid overwhelming the viewer. Those

All RNA Produced in the Nucleus Moves to the Cytoplasm

Time after incorporation of radioactive marker for Strain A1290	Total grains of marker incorporated into RNA	
	In Nucleus	In Cytoplasm
5-6 min	55.0	0
10 min	83.4	0
10 + 10 min	74.1	0
10 + 30 min	53.9	17.8
10 + 60 min	53.1	52.0
10 + 2 ½ hours	32.1	100.1
10 + 4 hours	22.4	108.9

FIGURE 14. Revision of figure 13 for a PowerPoint presentation.

interested in full details can and should consult the full scientific article or technical report. Turning a scientific article or technical report into a PowerPoint presentation is not just a shift in medium; it is a transformation from one medium to another. A PowerPoint presentation is not a substitute for a scientific article or technical report.

Graphs

As with tables, reading scientific graphs requires scanning, matching, and pattern recognition. In graphs, we scan to a particular data point and match it to a position on the abscissa (x-axis) and ordinate (y-axis). The result is a sentence for each data point in the form: "At time x, the magnitude of y is z." Pattern recognition involves judging whether a series of data points (or curves fit to these points) is increasing, decreasing, steady, or fluctuating in a predicable way. It might also involve a visual estimate of how close the data points are to the curves.

In Taylor's article the data "shown in Table 1 [are] interpreted in Figure 4"—

that is, the patterns of central tendency, obscured somewhat in the table by thickets of data and their standard deviations, are more clearly revealed in a figure. The two sets of points and their accompanying curves relate the progress over time of the number of grains in the cytoplasm and the nucleus. For the graph to be understood, the interpretation of its two curves must be integrated into a single message. When they are, the following sentence emerges: As the number of grains in the nucleus diminishes (curve descending), the number of grains in the cytoplasm increases (curve ascending). The proximity of the data points to the curves drawn through them suggests a strong experimental base for the tendencies described. The dashed line appended to the nucleus curve and departing from the data points reflects a correction applied as a result of measurements separate from those recorded in Taylor's table 1.

Figure 15 is Taylor's graph turned into a PowerPoint slide. Taylor's graph is not overburdened with data points or curves; each of the two curves has a identifying label. To incorporate the graph into a slide, we made only one change: we deleted the figure caption and inserted above the image a title abstracted from it. Perhaps we should also have used different colors for the

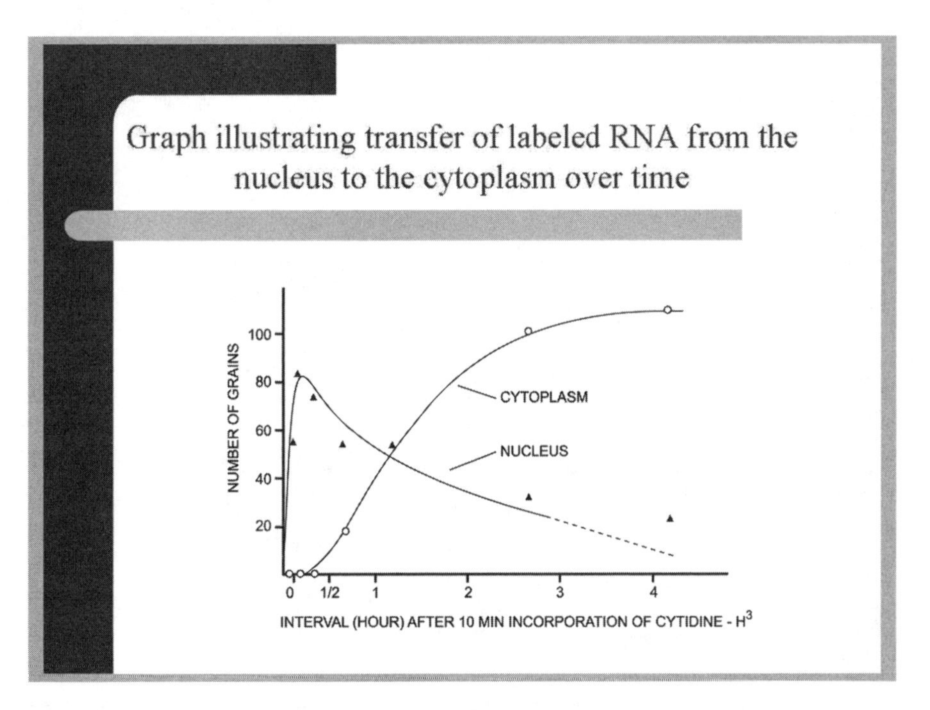

FIGURE 15. PowerPoint slide incorporating a graph. Graph reproduced with permission of Blackwell Publishing (Taylor 1960).

two curves and two sets of data points; perhaps we should also have added a label pointing to the dashed line and noting that it is derived from a correction factor. We would not change more for an audience of molecular biologists attending a conference.

Illustrations

In science, PowerPoint illustrations typically are photographs or drawings of the research objects under scrutiny or of the equipment used to investigate these objects. They are also interpreted by scanning, matching, and pattern recognition. We have three guidelines for reducing the mental processing required of such slides:

1. Choose illustrations that foreground their scientifically salient features.
2. Clearly indicate those features with labels, arrows, or circles.
3. For the labels, arrows, or circles, choose a font size and color that will stand out without being so conspicuous as to be a distraction.

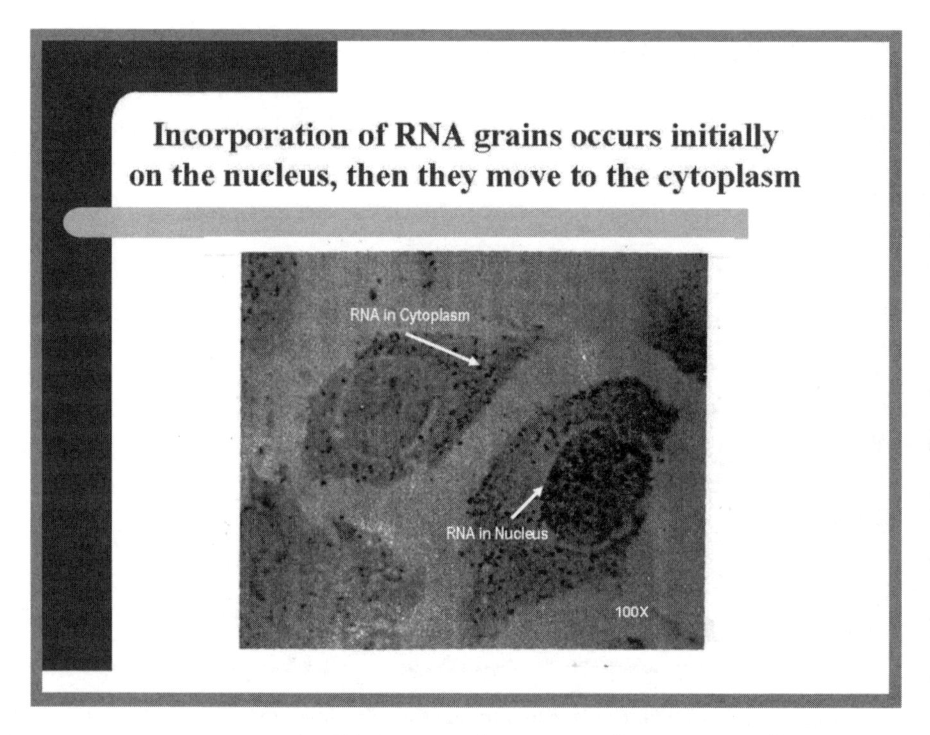

FIGURE 16. PowerPoint slide incorporating an autoradiogram. Autoradiogram reproduced with permission of Blackwell Publishing (Taylor 1960).

As an example, we have incorporated an autoradiogram from the Taylor article into a PowerPoint slide (fig. 16). The little black dots represent grains of cytidine-tritium incorporated into RNA. We added the title, descriptive labels, and arrows.

Lists

Another commonplace type of PowerPoint slide is the bulleted or numbered list. Few PowerPoint presentations, no matter what the subject matter, lack them. We have formulated the following PowerPoint list guidelines, a consensus based on the extensive literature on the subject:

1. Fewer than fifty words per list should be the norm; about twice that many should be the maximum.
2. Complete sentences should be the norm.
3. Parallel points should be in parallel syntax.
4. No more than two levels of hierarchy should be employed.

To illustrate, we created a bulleted list from Taylor's summary text, one that intentionally violates all four guidelines to the point of unintelligibility:

- Tritium-labeled nucleosides in hamster cells were grown in sterile culture.
- Cyclic changes of RNA were tracked in relation to cell division by means of audiograms.
  - Chromosomes
    - Of a single complement duplicated asynchronously.
    - Differences in timing among the different chromosomes and within a single chromosome.
  - RNA synthesis
    - Confined to the cell nucleus.
    - Cells in a medium free of labeled nucleosides. The RNA transferred to the cytoplasm after about four hours.

We now revise in keeping with the guidelines:

SEVERAL CONCLUSIONS WERE REACHED FROM AUDIOGRAMS OF TRITIUM-LABELED NUCLEOSIDES IN HAMSTER CELLS GROWN IN CULTURE.

- Chromosomes of a single complement duplicated asynchronously.
- Differences in timing occurred among the different chromosomes and within a single chromosome.
- RNA synthesis was confined to the cell nucleus.
- About four hours after the cells had been placed in a medium free of labeled nucleosides, the RNA transferred to the cytoplasm.

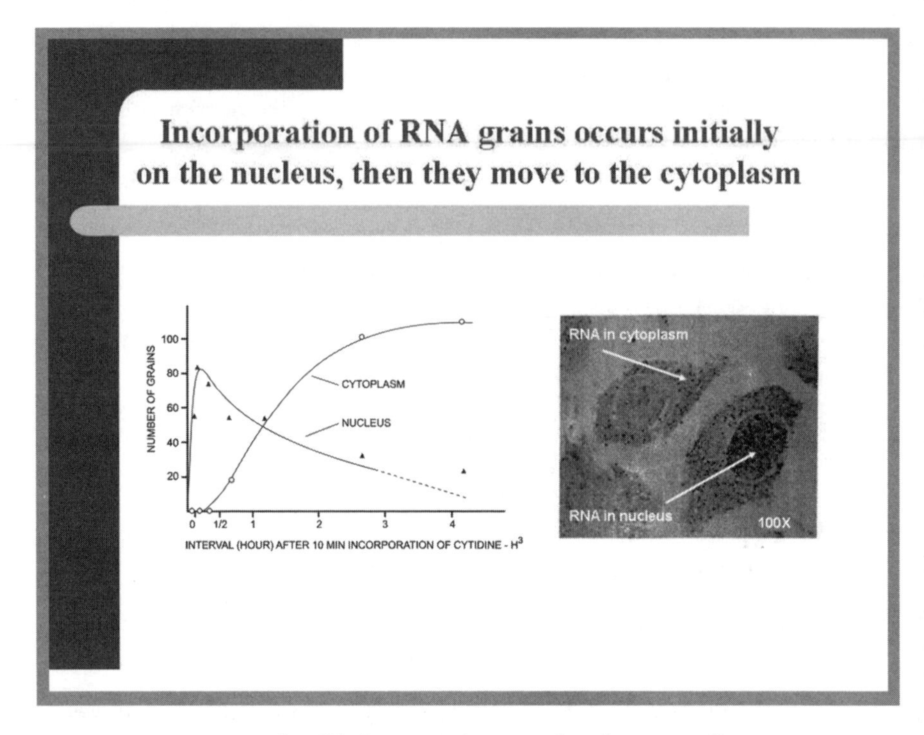

FIGURE 17. PowerPoint slide incorporating a graph and an autoradiogram. Images reproduced with permission of Blackwell Publishing (Taylor 1960).

A note of caution. In our view, good scientific presentations include relatively few slides displaying bulleted or numbered lists. Under the pressure to perform before an audience, the tendency is to just read the slide contents of such lists. Reading a series of such slides makes for a very dull talk. Good speakers use each item as a prompt, not a crutch. And good presentations greatly emphasize the visual over the written word.

Combinations

Many PowerPoint presentations combine two or more of the above elements on one slide. Nothing is wrong with that strategy, so long as all elements work together to deliver the same message, captured in the slide title. Figure 17 combines Taylor's graph and autoradiogram. Taylor's graph visually represents trends in the background-corrected data for strain A1290; Taylor's photo shows two biological cells from strain A1290, one with most of the RNA grains in the cytoplasm, the other with most in the nucleus. Both sides present strong visual evidence for the claim in the title.

Conclusion

There are many fine PowerPoint guides offering excellent advice on color schemes, background, font color and size, and numbers of objects per slide—that is, the aesthetic side of slide design. Our only advice is that you consider the audience. For a typical scientific meeting, PowerPoint presentations are versions of a scientific article in progress, with many of the technical details omitted. From a purely audience perspective, slides need only be readable: font large enough and color scheme not distracting. Preparing a review of your research for your research sponsors, you might behave differently; you might want to make the review as attractive as possible—scientifically and aesthetically. One dresses differently for a job interview than for a typical day at the lab or field.

EXERCISE

Scan the Web for slides and slide presentations and run through our checklist with a few that seem to have problems. Define these. How would you eliminate them? For example, what's wrong with the list in figure 18?

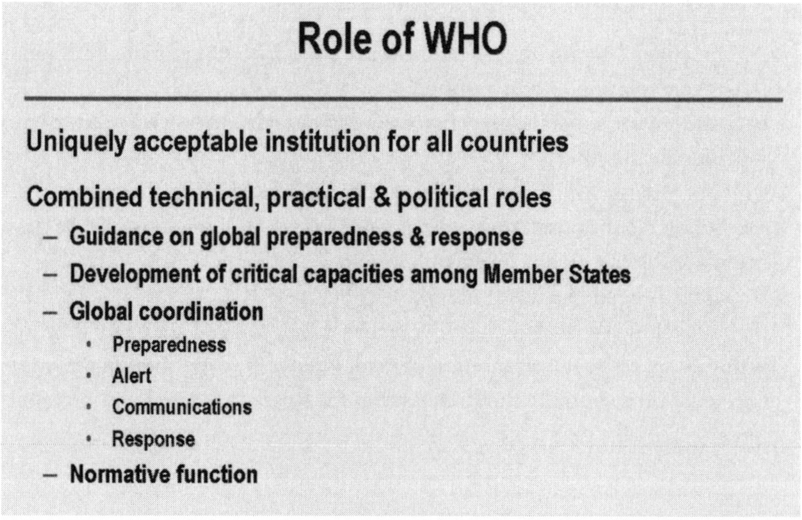

Role of WHO

Uniquely acceptable institution for all countries

Combined technical, practical & political roles

- **Guidance on global preparedness & response**
- **Development of critical capacities among Member States**
- **Global coordination**
  - **Preparedness**
  - **Alert**
  - **Communications**
  - **Response**
- **Normative function**

FIGURE 18. A slide in need of improvement.

Answer

This slide's three levels of hierarchy do not correspond to any logical relationship apparent among the points it makes. Our revision (fig. 19) attempts to clarify these relationships by employing complete sentences and parallel syntax. It also reduces the levels of hierarchy from three to two. Since there are two main topics, you might even want to divide them between two slides.

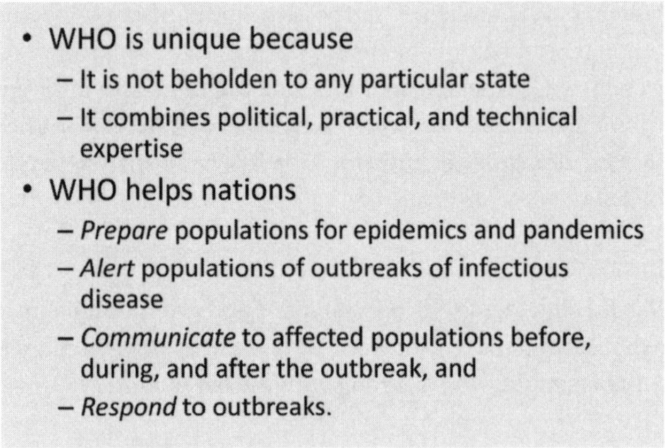

- WHO is unique because
  - It is not beholden to any particular state
  - It combines political, practical, and technical expertise
- WHO helps nations
  - *Prepare* populations for epidemics and pandemics
  - *Alert* populations of outbreaks of infectious disease
  - *Communicate* to affected populations before, during, and after the outbreak, and
  - *Respond* to outbreaks.

FIGURE 19. An improved slide.

CHECKLIST

Here is our checklist for tailoring your individual PowerPoint slides to reach an audience of auditors as distinct from readers:

- Do the slides in your presentation limit the points they make and make clear their relation to one another?
- In a slide with several objects, are all of your visuals keyed to the topic specified in the title?
- Do your slides avoid visual clutter and exhibit good visual contrast?
- Are bulleted slides used only when appropriate, and, when used, are they normally limited to two levels of hierarchy?
- Do your bulleted lists favor complete sentences?
- Is the level of visual and cognitive processing required for each slide appropriate to an oral presentation of visual material, even though the audience may be essentially the same as that for a research article on the same topic?

13 Organizing PowerPoint Slides

What comes out of PowerPoint depends largely on what goes into it; and the tool will likely neither improve poor thinking nor corrupt sound reasoning.

Jean-luc Doumont (2005)

PowerPoint presents the user with a problem of design: the design of each slide, a subject we covered in chapter 12, and the overarching design of a large set of slides, a subject we are about to cover. To help with this latter task, we offer five guidelines:

1. Begin your presentation with a slide that contains your title, your name, and, if available, a picture that visually reinforces the title's message.
2. Follow the title slide with a slide or slides introducing a research or societal problem.
3. Structure the main body of your presentation around three or four sequences of slides, each covering a different but logically related point.
4. Provide transitions between your sequences.
5. End your presentation with a slide or slides that reiterate the main points and, if appropriate, recommend future actions.

In our view, the best way to illustrate these guidelines is to analyze model PowerPoint presentations. We will look at three such models, two for a general audience and one for a professional audience. The first is "The Ongoing, Mind-Blowing Eruption of Mount St. Helens." It is by Dan Dzurisin (2006) of the Cascades Volcano Observatory, U.S. Geological Survey, and is an example of a PowerPoint talk for "science enthusiasts," that is, just about anyone with an interest in volcanoes. Dzurisin presupposes no highly specialized expertise in volcanology. The second is "House Hunting by Honey Bees: A Study in Group Decision-Making." Thomas D. Seeley (2005) of the Department of Neurobiology and Behavior, Cornell University, took this slideshow on the road before an audience of science students and college professors. Finally, we will analyze a professional-audience presentation, "A Dynamic Model of

Seismogenic Volcanic Extrusion, Mount St. Helens, 2004–2005" by Richard Iverson (2006), also of the Cascades Volcanic Observatory. It presupposes an audience of experts already thoroughly familiar with the science and graphic conventions in play and able to process complex visual information with great rapidity.

In this chapter, we will summarize the overall structure of each presentation and discuss the structure of selected slides. As you will see, as the intended audience changes from the general public to fellow experts, the mental processing of visual information required for the typical slide escalates.

(All the slides reproduced in this chapter employed color in the original PowerPoint versions. For economic reasons, we are able to reproduce them here only in gray scale. Some of their visual impact does get lost in the transition. We have thus set up a Web site for viewing all three presentations in their entirety as originally designed: www.press.uchicago.edu/books/harmon. Visitors can also go there to view each presenter's use of PowerPoint's animation feature.)

PowerPoint Aimed at Science Enthusiasts

Dan Dzurisin has an amazing story to tell—the spectacular 2004–5 volcanic eruption of Mount St. Helens in the state of Washington. Within the span of twenty-two slides, he tells us how the eruption was first detected in September 2004, how it developed, how the public was kept informed of these developments, and, finally, how the geoscientists kept themselves informed. Here is the complete narrative of the initial episode (in columns from left to right, we give slide number, slide title, and slide contents).

Slide 1	The ongoing, mind-blowing eruption on Mount St. Helens	[Photograph of erupting volcano from a distance, plus title, name of presenter, etc.]
Slide 2	Let's skip right to the good stuff . . .	[Photograph the lava dome the eruption created]
Slide 3	It started on a quiet September morning . . . with an earthquake swarm . . . that did not stop!	[Seismographs from two days as earthquakes began]
Slide 4	Within a few days, several earthquakes were occurring *per minute* . . .	[Seismographs for first week of seismic activity]
Slide 5	. . . and a large welt was rising on the south crater floor. Eight days after the first earthquakes . . .	[Photograph of welt the eruption created]
Slide 6	Mount St. Helens's first eruption of the twenty-first century was under way!	[Photograph of the eruption in action]

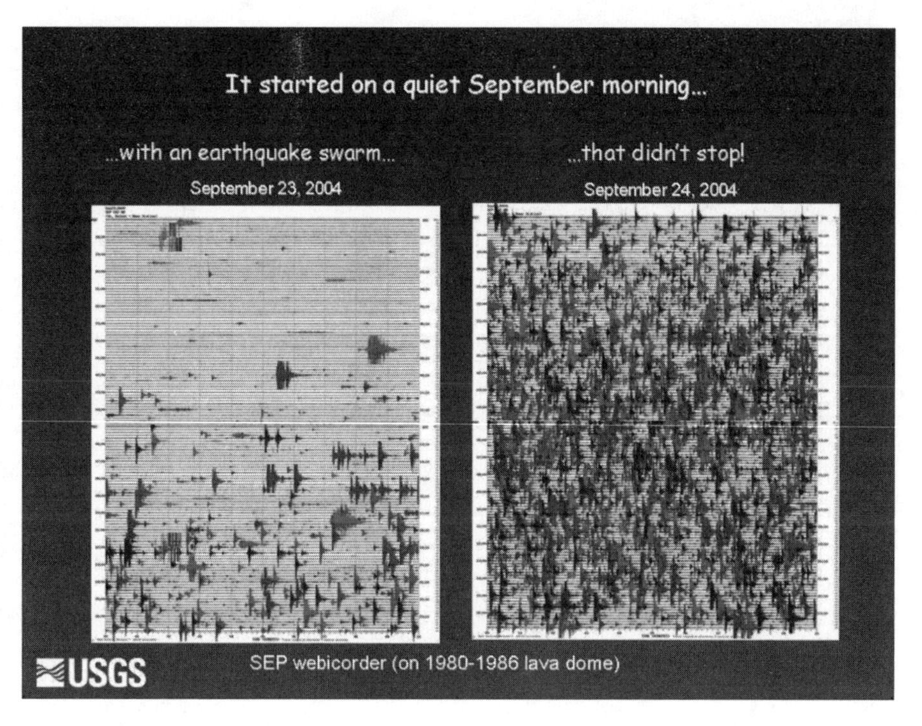

FIGURE 20. The eruption begins. Reproduced with permission of author (Dzurisin 2006).

From this introductory narrative, we immediately see that this talk is *not* about solving a new research problem of interest to specialists working on similar problems. Instead, it is about telling a story intended to grab the attention of science enthusiasts. Dzurisin begins his story with "the good [visual] stuff": a photograph that focuses on a new lava dome formed by Mount St. Helens's eruption (slide 2, not reproduced here). Like all good slides, this one conveys a single thought: this is what the volcano did. It provokes the question "How was this lava dome created?" That story unfolds in the subsequent slides in this initial sequence, linked one to the next through the slide titles.

We reproduce the third slide as figure 20. It exemplifies the close interweaving of the components of an individual slide, as well as the linking of one slide to others. This slide has one thought, an integral part of an overarching story, expressed in a short complete sentence at the top: "It started on a quiet September morning . . . " where the pronoun *it* refers to the pictures in the preceding slides and those that follow. The two graphs in the slide illustrate the "earthquake swarm" mentioned in the title, the cause of the lava dome just pictured.

In a sense, this third slide is really two slides. At first the audience sees only its left-hand side; next it sees both sides, experiencing thereby the dramatic contrast geologists experienced when monitoring the surface electronic properties (SEP) over two days in September 2004. Unless the speaker pointed out the character of the earthquakes depicted in the left-hand graph, however, the audience might not have noticed it: shallow, short lived, and steady. These are the "drumbeat" earthquakes that climax in the right-hand graph. They will form a component of a causal model of the eruption, the subject of Richard Iverson's professional talk, analyzed later in this chapter.

The next sequence of slides begins a shift from the science to keeping the public informed:

Slide 7	USGS [U.S. Geological Survey] and PNSN [Pacific Northwest Seismic Network] provided timely information and hazard assessments to partner agencies and the public.	[Plot of seismic activity from September to January]
Slide 8	The eruption sparked intense interest from the media and the public.	[Photograph of TV news vehicles parked at foot of mountain]
Slide 9	USGS scientists provided daily press briefings at CVO [Cascade Volcanic Observatory] until a joint information center was set up at USFS [United States Forest Service] Gifford Pinchot National Forest Headquarters . . . CVO kept its attention focused on the volcano . . .	[Photograph of USGS news conference]

We reproduce the first slide in this sequence as figure 21. It provides a transition from geological science to public information message. Its title announces the new topic: the communication of the seismic results gathered by the USGS for the media and public. The accompanying graph continues the story of the geological science and is linked to the previous slides by the box on the left with the label "Notice of volcanic unrest." The labels in the inset graph on the right link the science to making the science public.

The full graph may seem a tad overcrowded for public viewing and understanding. The audience of the presentation, however, sees the slide unfold in six manageable stages. First, they see only the slide title and the trace of seismic activity spanning the time from September 22 through January 24, 2005. The graph plots real-time seismic amplitude (RSAM) as a function of Pacific

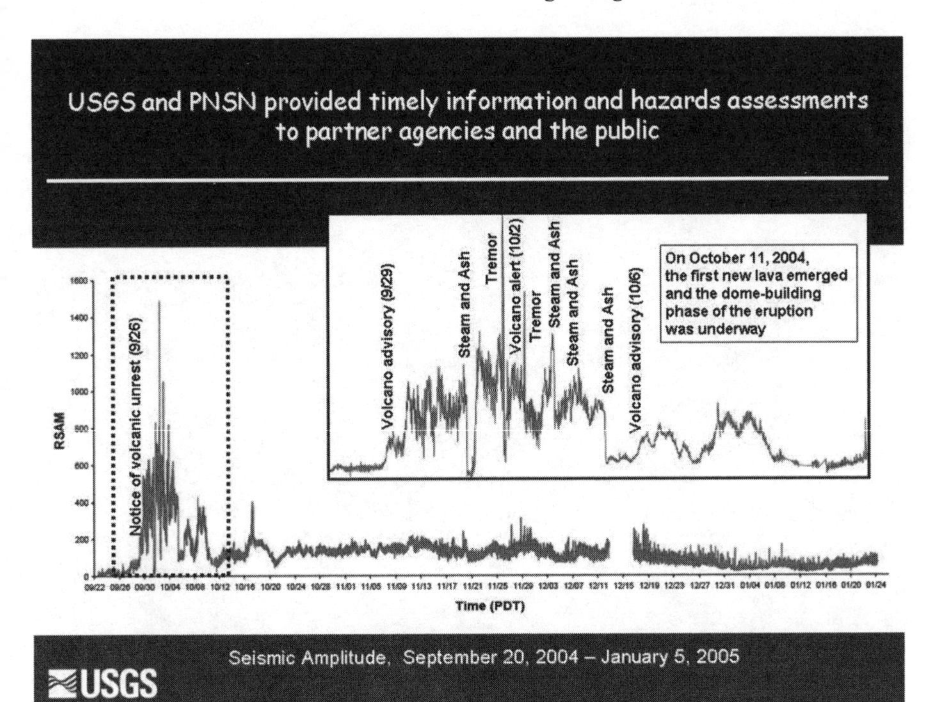

FIGURE 21. Graphing the eruption. Reproduced with permission of author (Dzurisin 2006).

Daylight Time (PDT). The boxes on the left and right are not yet visible. This first stage clearly shows major volcanic activity, followed by a gradual quieting over time. It presents the big picture. In the second stage, a box appears on the left of the screen with the words "Notice of volcanic unrest (9/26)." Here the speaker focuses our attention on the major volcanic activity from September 29 through October 11. After that, the focus turns to the box on the upper right of the screen: it amplifies a segment of the box on the left with labels added that interpret the meaning of the changing curve. The viewer sees the text for each date separately, allowing the speaker to focus on one date at a time. As a consequence of this sequencing, the audience sees not a cluttered graphic but an unfolding story: thanks to the precise monitoring of seismic activity, USGS and PNSN were able to provide timely information on the eruption to the media and public.

At the very bottom of the last slide in this sequence, the one with a photograph of a news conference, appear the words "CVO kept its attention on the volcano . . . " That serves to segue to the next sequence, emphasizing the

FIGURE 22. A helicopter monitors the eruption. Reproduced with permission of author (Dzurisin 2006).

monitoring of volcanic activity. Now the extent of CVO surveillance unfolds. We reproduce the first slide in this sequence (no. 10, here fig. 22): "The welt grew out at a prodigious pace (That's a *BIG* helicopter)." The white-letter, high-contrast labeling focuses our attention on the event; the circle around the helicopter gathering data (not quite so noticeable as in the original because of loss of color, but circle appears under the white-letter label "1980–1986 dome") highlights its monitoring. The foregrounding of the helicopter provides a sense of the grand scale of the drama below.

The final sequence of three slides brings the presentation to a close. So far, the audience has seen no bulleted lists—mostly they have seen eye-catching photographs with explanatory text. To the concluding slide (fig. 23), however, Tufte's criticisms of bulleted lists might seem to apply. On the one hand, we do not have a bewildering hierarchical system; on the other hand, we do have around two hundred words, well above our guideline. In the actual viewing, the scientist-presenter avoids information overload by use of the animation feature in PowerPoint, where the bullets fade into the slide one by one: in effect, the audience thus sees not one but four slides, presented in sequence, one bullet at a time.

Still, the presenter could have conveyed the same message to a general audience with considerably fewer words and greater visual impact. In our revised slide (fig. 24), the verbal and the visual work together: simultaneously, the viewer reads and sees "world-wide monitoring." The viewer need only scan from the short sentence at the top to the one at the bottom, pausing along the way to admire the colorful satellite photograph of an active volcano in the background. In that scanning, the viewer reads only about thirty words. The additional information in the original slide need not go to waste, however: some or all of it could be easily discussed by the speaker when displaying this slide.

Dzurisin's "conclusions" slide does not end the presentation; two more appear. Both show breathtaking photographs of an erupting Mount St. Helens. The first mentions that, starting from the midnineteenth century, the volcano's quiescent periods appear to have shortened alarmingly. The words on the second and last slide inject some humor: "This is not THE END!" A

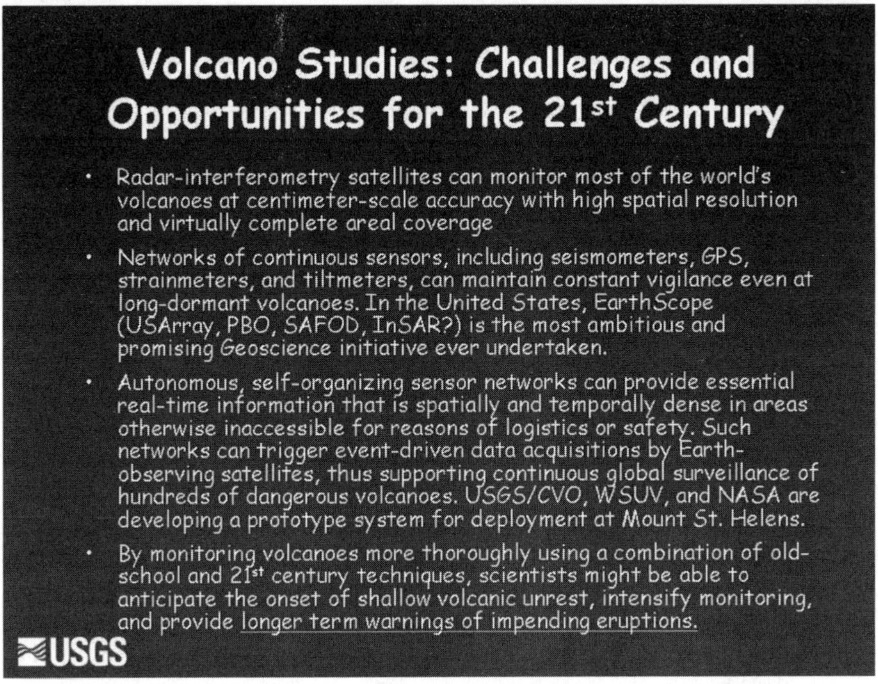

FIGURE 23. The concluding slide. Reproduced with permission of author (Dzurisin 2006).

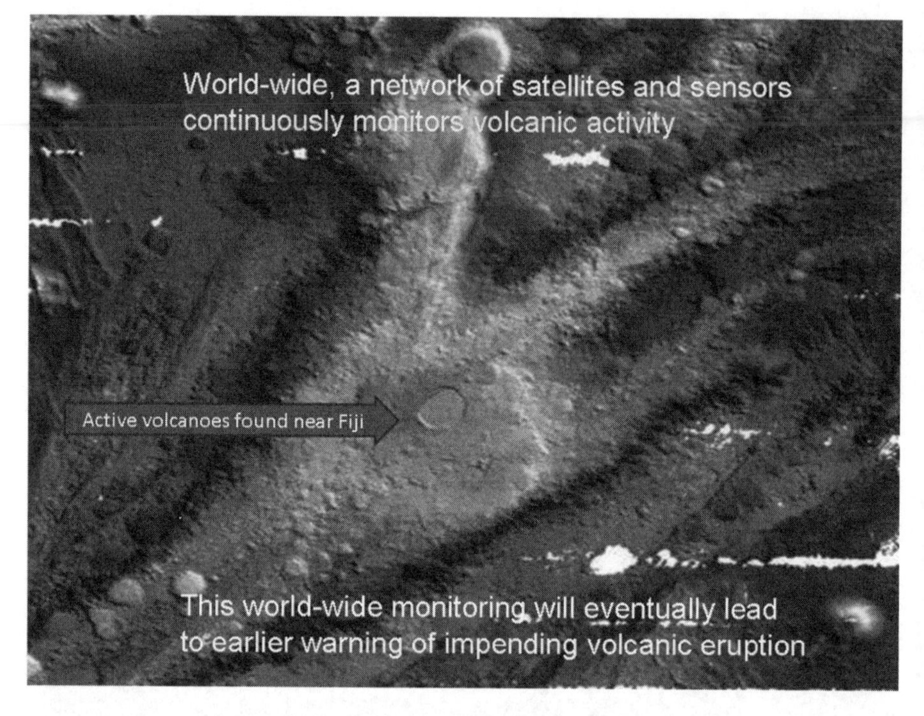

FIGURE 24. The concluding slide revised. Image reproduced with permission of Richard Arculus, Australian National University, 2008.

little humor at the end (or beginning) can go a long way when one is presenting PowerPoint science to an audience attending out of personal rather than professional interest.

Overall, Dzurisin creates a clear and compelling narrative that integrates the verbal and the visual and permits his general-public audience to reexperience what he and his fellow volcanologists experienced. Here are the key narrative elements:

- *Setting:* A volcano in the State of Washington
- *Main character:* A team of volcanologists
- *Plot:* The volcanologists detect the eruption, monitor its progress, and keep the public informed
- *Resolution:* The eruption is over, and the public has been kept informed and safe
- *Moral:* Public funds have been well spent in the public interest, and with continued funding, an even better story can eventually be told about active volcanoes worldwide

PowerPoint presentations aimed at the curious layperson tend to work best when one is telling a simple story with traditional narrative elements and displaying informative visuals that are pleasing to the eye, as Dzurisin does so well.

PowerPoint Aimed at College Students and Teachers

When we move from Dan Dzurisin's general-audience presentation of the Mount St. Helens eruption to Thomas Seeley's on decision making among honey bees, we shift from telling a story to addressing a series of research problems—the more typical job of scientific presentations. The expectations of the audience are somewhat higher in this case. We have shifted from a diverse group of science enthusiasts to an audience of college professors and their students.

In Seeley's presentation, the practical problem for the bees is this. A swarm has grown too numerous for its hive and must choose a new site to colonize. Scout bees, making up about 5 percent of the hive's population, have the job of finding the new home. Just after the title slide, Seeley presents the central question that motivates his research: "How can a group use the knowledge and opinions possessed by its members to produce an optimal choice of action for whole?" The slide is organized into two vertical columns (fig. 25). In the first, the question is posed; in the second, it is illustrated. The illustration is right on point: a visual presentation of the dilemma central to Seeley's research: How many hands are up? Must all hands be up before a positive decision is reached? Or only a majority? Or only a plurality? Below the picture we see the three main stages in group decision making—our introduction to Seeley's solution. In the sequence of eight slides that follow, Seeley marks out the research territory from which his chosen problem arose. He thereby covers the three basic elements of the typical scientific introduction we describe in chapter 1.

We consider the next slide we reproduce (fig. 26) an example of graphical excellence as defined by Edward Tufte (1983), "the well-designed presentation of interesting data." In it Seeley schematizes this decision-making process for honey bees scouting for a new nest. The slide contains a time series of eight graphs, which are read in the same order as one reads written text. These eight graphs represent vote totals, taken over a sixteen-hour period, for eleven potential new homes for the hive; the scouts vote by performing a dance in which they "waggle" in the direction of their choice. The angle of the waggle indicates the direction of a good prospect; the duration of the waggle indicates its distance.

While the slides that accompany presentations should form a coherent

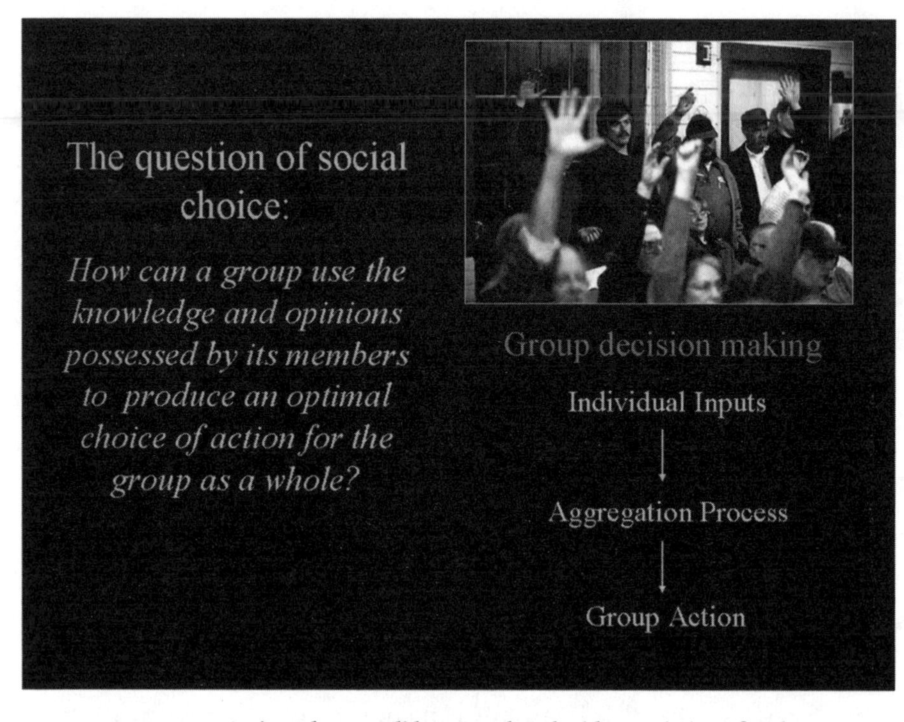

FIGURE 25. An introductory slide. Reproduced with permission of author (Seeley 2005).

set, they are meant to support, not to replace, the speaker. Speakers, not slides, are at the center of all oral presentations. With this particular slide, for example, the speaker must tell his audience how to read this graph. He must explain that the present bee hive appears as a circle with black dot in the center and that the arrows indicate the direction and relative distance of a potential new hive, as determined from the waggle dance. He must point out that each arrow has a letter code indicating a potential home, plus a number indicating how many scout bees "voted" in favor of it. The thickness of each arrow, he must add, correlates with the number of votes received by a given site.

Decoding this graph in detail takes what would appear to be a considerable amount of scanning, matching, and pattern recognition. Yet because of the ingenuity of the slide's design, once the speaker has explained how to read the first graph in the time series, the viewers immediately have the key to the other seven. As a consequence, the slide's message comes through clearly and easily. Site G first appears in the second graph with four votes, but support for it grows steadily over time; as it does, its arrow grows in thickness, until, when it wins by a unanimous vote in the eighth slide, all competing arrows have vanished.

The slide also contains a useful visual device. Look at the graph on the upper right. You will see a kilometer scale and compass. They inform us that the new home (labeled "G") lies about two kilometers from the present site, in a southwestern direction. Once the scout bees have made their unanimous decision, the swarm heads for the new home.

Seeley has invented a visual language appropriate to the presentation of his field data. In this series of diagrams we actually see, actually experience, decision making in all of its complexity—a process that terminates in an "aha!" moment represented by the utter simplicity of the single arrow of the final panel. Those who want to study the graph more carefully can consult the journal article cited at the bottom right.

A subsequent slide (fig. 27) reinforces the point Seeley made in the graph. What is *symbolized* in the graphs is *depicted* in the drawing: as the election

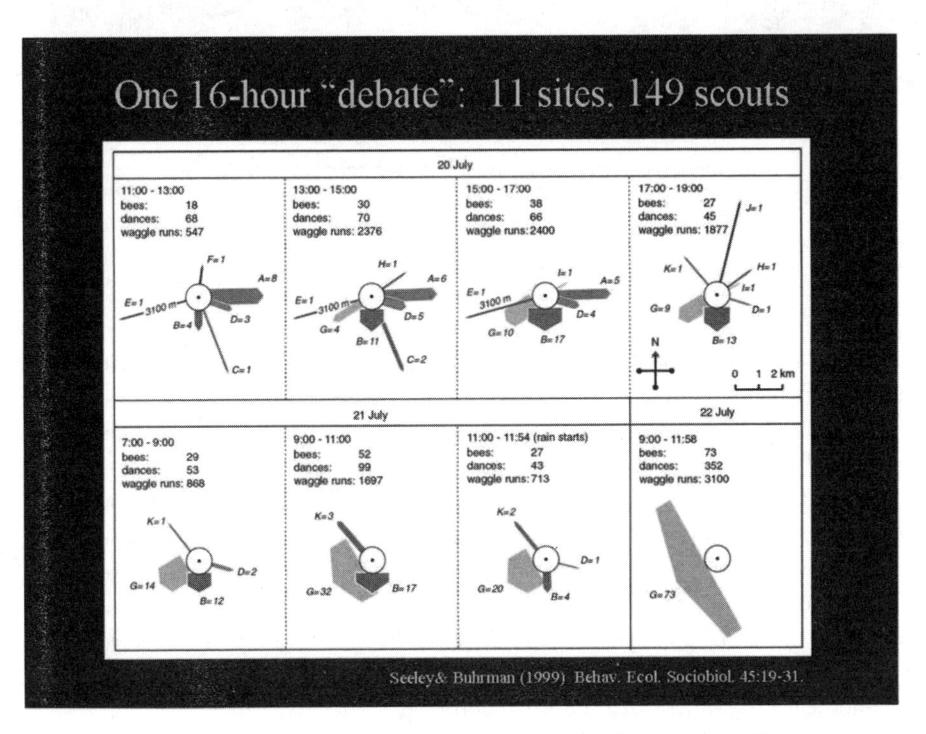

FIGURE 26. The swarm finds a new home: A graphic representation. Reproduced with permission of author (Seeley 2005).

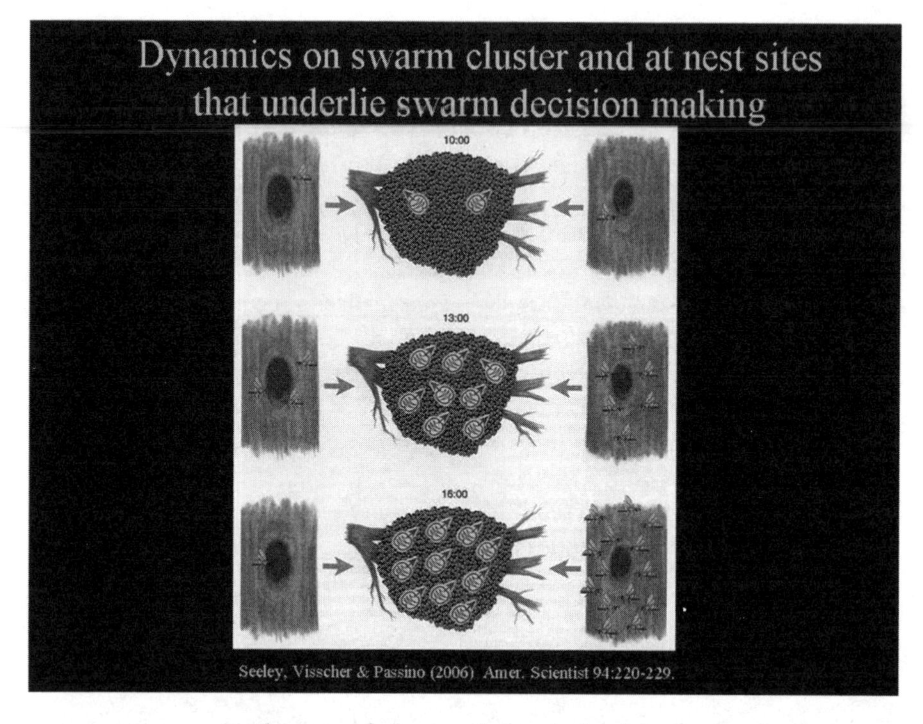

FIGURE 27. The swarm finds a new home: A pictorial representation. Reprinted with permission of author and *American Scientist* (Seeley 2005; Seeley, Visscher, and Passino 2006).

plays out, the scouts eventually concur and waggle toward the same new home, after which the bees emigrate en masse. As we discussed in chapter 5, good scientific communications continually reinforce key messages without being overly repetitious.

All good PowerPoint presentations conclude with one or more slides devoted to the lessons the speaker wants the audience to take home with them. Seeley's conclusion appears in a sequence of three slides. They all address the introductory research question: can bees teach us anything of value for our own decision making? The first acknowledges the well-known problem of "groupthink" over individual creativity. In contrast, the next slide, reproduced here (fig. 28), uses a catchy title, numbered list, and photo of a beehive to highlight the benefits of group decision making. It sums up for a general audience the three main lessons that the bees teach us. Supporting each lesson are one or two scientific observations about the scouting bees. In a clever turn, when the speaker presses forward with the slide show, the list remains as is but the bees in the upper right corner disappear. A human brain magically appears in their place, dramatizing the analogy between the swarm and us.

Storytelling is perhaps the best strategy when we are dealing with a general audience of uncertain scientific background. But in addressing his audience of college students and professors Seeley does not tell a story; he does not relate a series of unique events that form a coherent whole, as in Dzurisin's case. Rather, he reveals the details of a process that in all its essentials repeats itself over time: the way swarms of bees reach a decision about the location of a new hive. He claims that they reach this decision by means of a quorum, one that is far more often right than wrong. The data he amasses and the visuals he displays support this claim, a claim he extends to human groups. Seeley brings his argument home to his audience by showing how his conclusions apply to their own lives. Seeley's research was time consuming and arduous; it required patient attention over long periods and sedulous attention to detail. In this PowerPoint presentation, however, his argument has a structure that is simplicity itself.

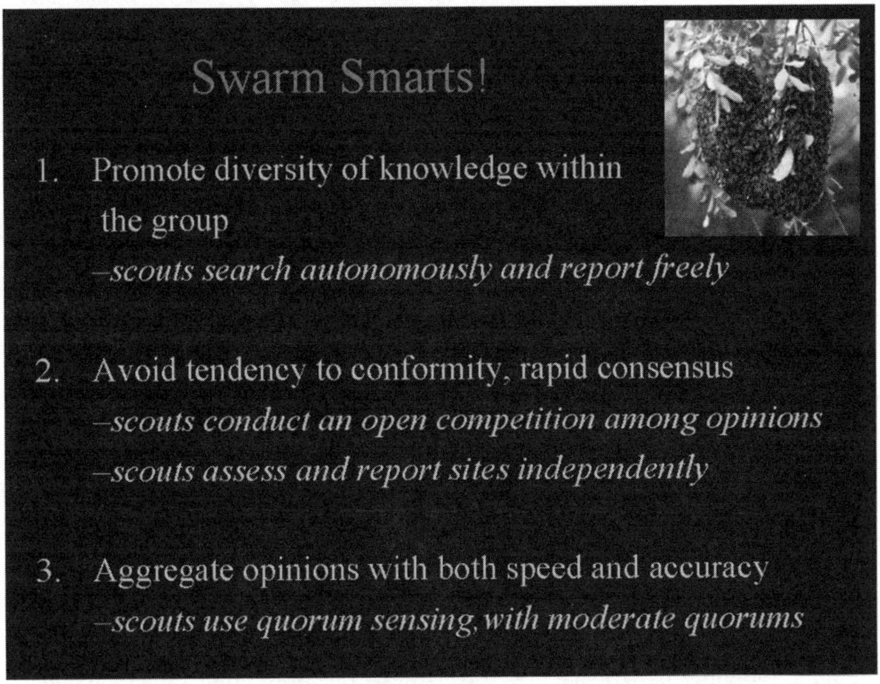

FIGURE 28. A concluding slide. Reproduced with permission of author (Seeley 2005).

He begins with an introductory slide posing a provocative question in whose answer his learned audience might be interested. This is followed by a sequence of slides on the basics of bee behavior. We then see several slide sequences, all revolving around the answers to four specific research questions:

- Do the scouting bees reach a decision by consensus or by quorum?
- How is the decision of the scouts conveyed to the rest of the swarm?
- Is that decision the best choice?
- Can the bees teach us anything about our own decision making?

The three concluding slides summarize and generalize his findings.

No matter how complicated the research presented, good PowerPoint presentations employ a simple organization that any informed person can follow. We strongly recommend sketching out such a structure on paper before starting a new PowerPoint presentation, then revising the structure as your presentation develops in further drafts.

We close with a note regarding the bookends to Seeley's presentation: the title slide and an acknowledgments slide. Both are typical for presentations reporting new research. On the initial slide are the presentation's title emphasizing Seeley's subject, "house hunting," a photo of a beehive illustrating his object of study, and his name and affiliation. There is no mention of anyone but the presenter, though many others participated in the research. As is standard practice, Seeley leaves it to the very last slide to acknowledge others—seven research partners, five field assistants, an inspirational mentor, and three funding sources. (You might want to consult our chapter 7 for additional guidance on giving credit.)

PowerPoint Model Aimed at Professional Audience

Earlier we discussed a PowerPoint presentation by Dan Dzurisin on the Mount St. Helens eruption in 2004–5. With Richard Iverson's presentation on the same topic, we shift from a general to a professional audience. This shift is clearly signaled by the titles on the first slides: "The Ongoing, Mind-Blowing Eruption of Mount St. Helens" for the former; "A Dynamic Model of Seismogenic Volcanic Extrusion, Mount St. Helens, 2004–2005" for the latter. Note the underscored nucleus nouns in the titles (see chapter 3 on titles). They tell us that Dzurisin will be describing an eruption for those interested in a "mind-blowing" experience, while Iverson will be presenting a physical and mathematical model of the eruption.

Iverson's central task is to present a model that explains three facts about the eruption.

Fact 1. A large plug of volcanic rock formed, fitting into the volcano cone like a loose-fitting cork in a bottle. Because of the powerful turbulence below, for about a year this plug had moved up and down at a nearly constant rate.

Fact 2. The newly formed extrusions created a series of spines whose freshly exposed surfaces exhibited a striated fault gouge formed by the movement of the plug against the sides of the cone. Analysis of this gouge showed that the rate of movement of the plug had weakened over time.

Fact 3. In the area of the volcano, repetitive "drumbeat" earthquakes caused by the oscillation of the plug occurred about every hundred seconds, had a magnitude of less than two, and were centered at depths of less than a kilometer, directly beneath the plug the eruption created.

Iverson's introductory sequence of six slides establishes, comments on, and visually represents different aspects of these three facts. In our discussion, we will focus only on fact 2. In the first slide we reproduce (no. 4; our fig. 29), we see with our own eyes what soon will be captured by a mathemati-

FIGURE 29. Striated fault gouge: A photograph. Reproduced with permission of author (Iverson 2006).

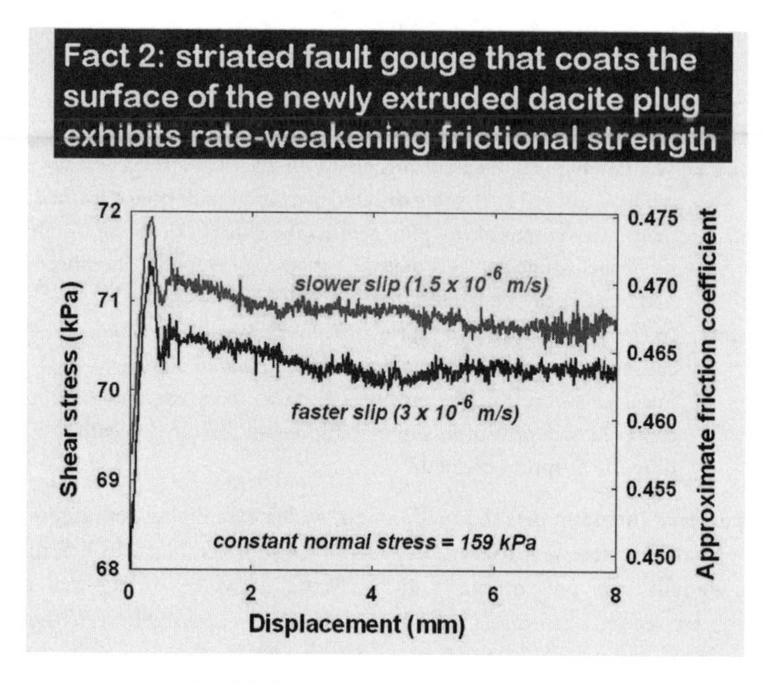

FIGURE 30. Striated fault gouge: A graph. Reproduced with permission of author (Iverson 2006).

cal model: the striated fault gouge at the heart of fact 2. The arrow zeroes in on the gouge. The slide conveys a single thought, verbally and visually: this is what the volcano did. The slide provokes the research question "How was this striated fault gouge created?" Even in this fairly simple slide, the technical language in the text box that serves as the heading clearly assumes a professional audience.

The next slide (fig. 30) repeats the same text box, but the accompanying image is a graph analyzing the event. The presence of the same heading binds these two visuals into a single cognitive unit. The results indicate that slip, the upward motion of the volcanic plug photographed in the previous slide, increases as a consequence of reduced friction. In the graph, shear stress, stress parallel to the face of the material, is measured in kilo-pascals (kPa), ten of which equal one atmosphere of pressure. The graph shows two curves: a red one for a slower slip and a blue one for a faster slip. The different colors facilitate discrimination and comparison. Taken together, the two slides give us a rounded picture of the "fact" expressed in the heading: we see the gouge as it exists in nature, and we see its transformation into standard physics measurements.

What accounts for fact 2? The authors model this volcanic system as a damped oscillator, one whose back-and-forth movements are reduced over time. These movements are characteristic: they are called "stick-slip," repetitive intermittent displacements caused by changing frictional force. It is this motion, this oscillation, that causes the repetitive "drumbeat" earthquakes mentioned in fact 3. The next slide we reproduce (fig. 31) shows a mechanical model of this oscillation. A plug of mass *m* under pressure from the rising molten rock or magma *p*, and counteracted by the force of friction *F*, oscillates around a point of equilibrium as a consequence of the drumbeat earthquakes, which diminish in intensity over time. Beneath each variable, parameter, and constant in the diagram appear its circled shorthand symbols. These circles are color coded for ease of identification: brown for constants, blue for parameters, and red for variables. Color coding (not visible in the reproduction) is an effective means of singling out the separate factors that combine to cause the eruption. (SPASM stands for seismogenic play of ascending, solidifying magma; 1-D stands for one dimensional.)

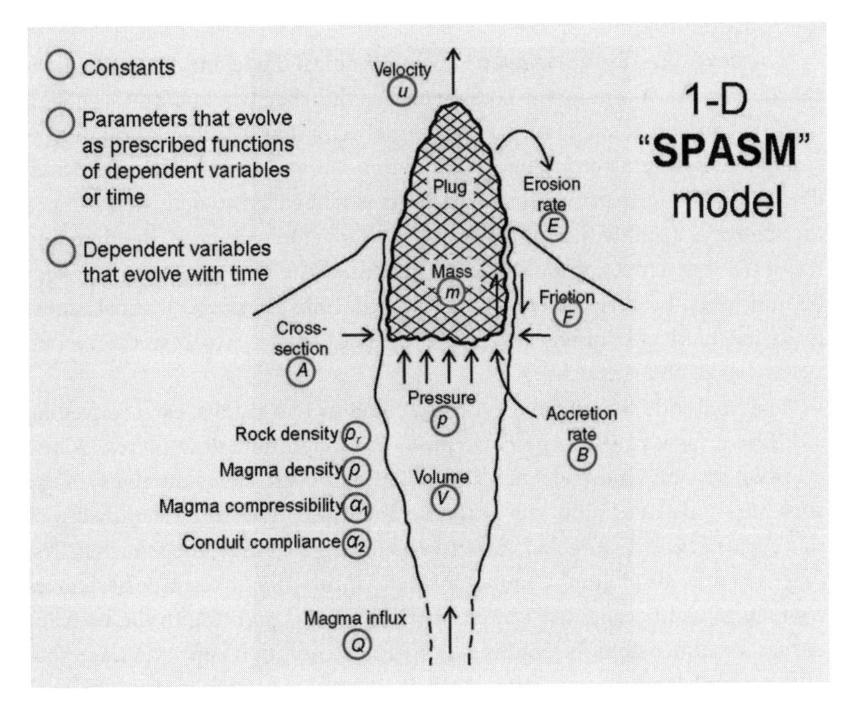

FIGURE 31. Volcanic eruption: A mechanical model. Reproduced with permission of author (Iverson 2006).

If $\kappa = 0$, $B = Q$, and t_0 is constant, behavior of numerical solutions depends almost entirely on D evaluated at the equilibrium slip rate $u = u_0 = Q/A$:

$$D = \frac{1}{2}\frac{t_0}{m_0}\left[\frac{dF}{du}\right]_{u=u_0} = \frac{1}{2}\left(c\lambda\mu_0 \frac{g t_0}{u_0}\frac{u_0}{u_{ref}}\left[1+\left(\frac{u_0}{u_{ref}}\right)^2\right]^{-1/2}\right)$$

which simplifies to

$$D \approx \frac{1}{2}\left(c\lambda\mu_0 \frac{g t_0}{u_0}\right) \qquad \text{if } u_0/u_{ref} >> 1$$

FIGURE 32. Volcanic eruption: A mathematical model. Reproduced with permission of author (Iverson 2006).

The next slide (fig. 32) converts this mechanical model into its mathematical counterpart. Although the mathematics of this slide would not sit well with a general audience, such fairly conventional equations would not faze professionals attending a conference presentation about a new model of volcano dynamics. Typically, this set of equations is embedded in a single sentence that conveys a single thought: when the plug velocity is high, the damping factor (*D*) is to a first approximation a function of the rate-weakening strength (*c*) multiplied by that velocity. This equation links the mathematical model to its mechanical counterpart. *D* in the equation matches *D* in the model, *u* matches *u*, and *F* matches *F*.

The final slide we reproduce (fig. 33) contains two graphs, each conveying a different facet of the same concept: as the magnitude of *D* increases, the eruption becomes more violent. The graph on the left shows the shape of the stick-slip cycle over time; the graph on the right shows the relationship of this cycle to the pressure and velocity. In viewing this slide, the audience first sees only the pair of graphs and their labels; at this juncture, a blue bar moves from stage right across the screen, its twin arrows pointing to the relevant curves for dimensionless damping, *D* = –2, the condition approximating that of the actual Mount St. Helens eruption. It is this blue bar that has created emphasis throughout Iverson's presentation.

In his last slide, Iverson lists the five conclusions he has derived by applying his theoretical model to eruption dynamics. He does not recommend future actions or discuss wider implications, in contrast to Dzurisin's presentation on the Mount St. Helens eruption. Creators of PowerPoint presentations for professional audiences should at least consider those two elements in any conclusion. However, as discussed in chapter 5, their presence is by no means obligatory.

Typically, such presentations as Iverson's also start with an introduction that establishes a research problem. Because this presentation does not have such an introduction, we have taken the liberty of constructing one by borrowing from Iverson's *Nature* article on the same subject. We have followed the principles of introductions presented in chapter 1:

- [*Problem in the field of volcanology:*] Volcanic eruptions are difficult to model.
- [*What is already known about one particular volcano:*] The recent eruption at Mount St. Helens exhibited near-equilibrium behavior over about a year.
- [*More on what is already known:*] This behavior includes nearly steady extrusion of volcanic material coupled with periodic shallow "drumbeat" earthquakes.

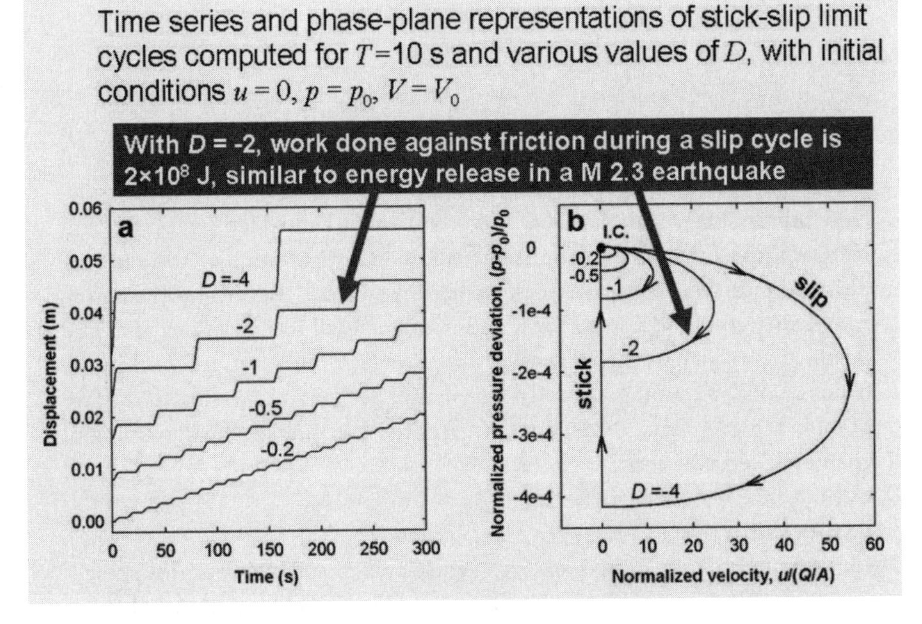

FIGURE 33. Graphs derived from model calculations: Mount St. Helens eruption at D = −2. Reproduced with permission of author (Iverson 2006).

- [*Assertion that the authors have found one possible solution to above problem. Continue listening to learn more:*] This behavior has been modeled mechanically and mathematically as a damped oscillator.

Iverson's professional presentation is like its general-audience counterpart in that it follows most of our PowerPoint guidelines in chapters 12 and 13. Highly technical content does not give the presenter a reason for bewildering the audience with irrelevant illustrations, visual clutter, verbosity, forests of bullet points, or sentence fragments whose connection to the whole remains mysterious. Overall, despite the complexity of the science, the logic of Iverson's presentation is straightforward—three sequences and a conclusion slide. Its structure is as simple as it can be, given the overriding need to make a definite claim and to support it with evidence.

The first sequence presents three facts gathered about the eruption (with a short movie thrown in for our entertainment and edification). The second sequence presents the mechanical and mathematical model created to explain these facts. The third sequence presents the calculations that establish the credibility of the mechanical model by deducing from its mathematical counterpart a set of consequences that match the data with a satisfying degree of approximation. Good PowerPoint presentations of this length tend to contain three or four such sequences. The simpler the overall structure, the better it is for presenting complex technical material.

Conclusion

In the struggle to maintain an audience's attention, PowerPoint is a useful though potentially dangerous tool. It is potentially dangerous because its default conditions—its six levels of hierarchy, other embedded templates, and many options for slide transitions and color—constitute an invitation to cognitive nightmare. In the worst presentations, verbal clutter and byzantine complexity combine to transform audience goodwill into bewilderment and hostility. But such abuse is unnecessary. PowerPoint is a useful tool in the hands of skilled communicators like Dzurisin, Seeley, and Iverson. In the face of their achievements, Professor Tufte's criticisms, cited at the beginning of chapter 12, do not apply.

EXERCISE

Choose a heavily illustrated research article and create a PowerPoint presentation from it. Imagine that the audience is scientifically knowledgeable but not experts. Can you make the title less technical and forbidding? Simplify the visual and cognitive processing required by the audience for any tables or figures incorporated? Create bulleted lists for the introductory and concluding slides? Create two or three discrete sequences of slides in between?

CHECKLIST

After you have finished the above PowerPoint exercise, ask yourself the following:

The Slides

1. Does the nucleus noun in your title emphasize the key aspect of the research?
2. Can you complement your title with a visual representation?
3. Did you consider the elements in scientific introductions and conclusion sections when creating your introductory and concluding slides?
4. Does each slide in your presentation limit the points it makes and at the same time make clear the relationship among its points?
5. Is the level of visual and cognitive processing required for each slide appropriate to auditors as distinct from readers?

The Sequence

6. Do you make good use of visual and verbal cues linking the slides in a sequence?
7. Does your presentation consist of a series of sequences, each with a clear relationship to the whole?
8. Are there clear transitions between your sequences?
9. Does your presentation have an overarching structure, with a central theme that each slide furthers, a theme about which the audience is reminded throughout?

PART III

Writing Style

14 Composing Scientific English

We are such stuff
As dreams are made on, and our little life
Is rounded with a sleep . . .

William Shakespeare

For humans are of the type of material that constitutes dreams, and their relatively brief existences terminate, as well as taking their inception, in a state of unconsciousness.

R. F. Gombrich (1989)

By translating Shakespeare's famous lines from the *Tempest* into stilted academic prose for readers' amusement, the British scholar Richard Gombrich exhibits his dissatisfaction with such prose. The same dissatisfaction has motivated many commentators to criticize scientific writing as having degenerated into a "molasses of jargon and academic code." We would guess many scientists would agree with that opinion, except as it might apply to their own prose. But we believe it is a facile generalization and not much help to anyone wanting to improve his or her scientific prose. Here is why.

Current scientific English as exemplified in the best journals differs markedly from the somewhat more reader-friendly prose that characterized scientific communication at the beginning of the scientific revolution in England. To illustrate this point, let's examine a seventeenth-century passage from Sir Robert Boyle's *Experiments and Considerations of Colours* (1666). We apologize for the length of this passage, but only at such length is the character of Boyle's style fairly exhibited:

> I know not whether I may not on this occasion add, that Colour is so far from being an Inherent quality of the object in the sense that is wont to be declar'd by the Schools, or even in the sense of some Modern Atomists,

> that, if we consider the matter more attentively, we shall see cause to suspect, if not to conclude, that though Light do more immediately affect the organ of sight, than do the bodies that send it thither, yet Light it self produces the sensation of a Colour, but as it produces such a determinate kind of local motion in some part of the brain; which, though it happen most commonly from the motion whereinto the slender strings of the *Retina* are put, by the appulse of Light, yet if the like motion happen to be produc'd by any other cause, wherein the Light concurrs not at all, a man shall think he sees the same Colour. For proof of this, I might put you in mind, that 'tis usual for dreaming men to think they see the Images that appear to them in their sleep, adorn'd some with this, and some with that lively Colour, whilst yet, both the curtains of their bed, and those of their eyes are close drawn. And I might add the confidence with which distracted persons do oftentimes, when they are awake, think, they see black fiends in places, where there is no black object in sight without them. But I will rather observe, that not only when a man receives a great stroak upon his eye, or a very great one upon some other part of his head, he is wont to see, as it were, flashes of lightning, and little vivid, but vanishing flames, though perhaps his eyes be shut: But the like apparitions may happen, when the motion proceeds not from something without, but from something within the body, provided the unwonted fumes that wander up and down in the head, or the propagated concussion of any internal part in the body, do cause about the inward extremities of the Optick Nerve, such a motion as is wont to be there produc'd, when the stroak of the Light upon the *Retina* makes us conclude, that we see either Light, or such and such a Colour.

Let us ignore small differences in spelling and punctuation and the ornate sentence structure and phrasing common to the period; they are not material to the point we want to make. Even discounting these, we can see that Boyle's prose lacks the characteristics we associate with current scientific English.

In the first place, Boyle does not use specialized technical terms except perhaps "Optick Nerve" and "Retina." A vast lexicon of scientific and technical terms did not yet exist. Second, Boyle does not favor verbs in the passive voice: for example, rather than the modern "if the matter *is considered* more attentively," Boyle writes "if we consider the matter more attentively." Moreover, Boyle exhibits no qualms about using first-person pronouns (*I* and *we*) repeatedly.

Third, nominalization practices differ greatly. (Nominalization is the grammatical alteration of verbs into nouns—the alteration, for example, of *evolve* into *evolution.*) Its systematic use is another characteristic of current scientific English. In Boyle's prose, however, nominalizations are few and far between: his commonplace choices, for example, "sensation" and "motion," pass without notice. Fourth, complex noun phrases, so characteristic of current scientific English, are scarce in Boyle. His average noun phrase is short—it is a noun and its accompanying article, adorned occasionally by an adjective and a prepositional phrase: for example, "the slender strings of the *Retina.*" Indeed, Boyle's short noun phrases are so numerous that the lengthiest of them—"the Inherent quality of the object in the sense that it is wont to be declar'd by the Schools"—passes almost without notice. Also entirely absent are noun strings, that is, a central noun preceded by multiple modifiers, as in "planar graphite fused six-membered ring structure" (Kroto et al. 1985), used to designate a new form of carbon. Finally, Boyle's prose differs in its use of verbs. A characteristic of current scientific English, a corollary of the omnipresent long, complex noun phrase, is the use of weak verbs and of the same few verbs repeatedly: for example, the predicates *to be, to have, to show, to find, to use.* In contrast, Boyle "considers," "concludes," "observes," "endeavors"; his men "see," "think," and "receive"; his light "produces," "concurs," "causes."

Clearly, over the centuries a sea change has occurred. Scientific English has evolved, just as standard written English has. In this chapter, we describe the general characteristics of the current state of this evolution: a fondness for verbs in the passive voice; the systematic use of nominalization, long noun phrases, and noun strings (and their concomitant weak verbs); and, finally, the proliferation of technical terms. We believe in the middle path to good scientific prose: that is, not avoiding these characteristics if at all possible, nor blindly using them at every turn of phrase, but mastering them.

A Strong Reliance on the Passive Voice

Typical verbs express an action—what he, she, or it did: "The boy hit the ball." The verb is in the active voice. But English also allows you to express the same thought differently: "The ball was hit by the boy." The verb is now in the passive voice. The agent of the action, the boy, appears after the verb, while the object of the action, the ball, appears before it. In the passive voice you can also omit the agent if you wish. You can write: "The ball was hit." The passive voice is a way that English allows writers to remove explicit mention of the human agent from a sentence. An object, process, or concept appears in the subject position.

Those dispensing how-to-write advice routinely advocate shunning the passive voice and embracing the active. Some writing teachers go so far as to see any use of the passive voice as a sign of weak writing. So construed, *passive* equals bad, *active* equals good. That formula is even reinforced by the choice of names for the two verb voices. In marked contrast, some scientists seem to believe that good scientist-writers ought never use the first-person pronoun *we* or *I*. Suppressing the first-person pronoun thus forces them to use passive voice verbs in abundance.

In reality, scientists have freely used both passive verbs and first-person pronouns from the beginning of modern science. Our own research has shown that the passive voice appeared on average about once per hundred words in the seventeenth century. That rate steadily increased over time until it had doubled by the early twentieth century. Thereafter, it has remained relatively flat. Still, with more than two uses per hundred words, modern scientists clearly do not practice the advice preached by many writing teachers. And given another twentieth-century norm, one first-person pronoun per hundred words, they do not forgo this modest "personal" touch either.

Particularly in methods sections, you will likely find English scientific prose heavily laden with verbs in the passive voice. To exemplify this practice, let us turn to a passive passage from David Baltimore's "Viral RNA-Dependent DNA Polymerase" (1970), a classic paper on understanding viruses and designing drugs to combat them:

> A preparation of R-MLV containing 150 μl. *was layered* over a linear 5.2 ml. gradient of 15–50 per cent sucrose in PBS-EDTA. After centrifugation for 2 h at 60,000 r.p.m. in the Spinco "SW65" rotor, 0.27 ml. fractions of the gradient *were collected* and 0.1 ml. portions of each fraction *were incubated* for 60 min in a standard reaction mixture. The acid-precipitable radioactivity *was then collected and counted.* The density of each fraction *was determined* from its refractive index. (emphasis ours)

Now let's change the original passive verbs into active:

> *We layered* a preparation of R-MLV containing 150 μl. over a linear 5.2 ml. gradient of 15–50 per cent sucrose in PBS-EDTA. After centrifuging this mixture for 2 h at 60,000 r.p.m. in the Spinco "SW65" rotor, *we collected* 0.27 ml. fractions of the gradient and incubated 0.1 ml. portions of each fraction for 60 min in a standard reaction mixture. *We then collected and counted* acid-precipitable radioactivity. *We determined* the density of each fraction from its refractive index.

Is one passage better than the other? Not really. They are just different. In the first, the performers of the different actions in the lab are hidden from view. In the second, the human agents take center stage. It comes down to a matter of emphasis. What is the central story here: what the authors did in the lab or what was done by them in the lab? Baltimore decided on the latter. Most scientists would agree with his choice.

Here is the important point about the selection of voice. When good writers want to stress or make unequivocal what they or some other person did, they begin sentences with the pronoun *we* (for a paper with multiple authors) or *I* (for a single author) or a person's name (Harmon, Gross, etc.). When they want to emphasize the objects of the world, the products of scientific methods and the laboratory, or other inanimate constructs, they frequently must use the passive voice. In general, current scientific English relies heavily on the passive voice because science is about objects and events, not people.

A Network of Nominalizations, Complex Noun Phrases, and Noun Strings

Current scientific English is characterized by nominalizations: nouns converted into verbs. Nominalizations often end with either "-tion" or "-ment," as in *development,* but sometimes they undergo no change in form:

> They pursued *research* on how amphibians evolved from fish.
> They *research* how amphibians evolved from fish.

You can find an abundance of nominalizations in almost every sentence of current scientific English. Take the following from an article by Wilmut and collaborators (1997) on the first mammal to be cloned from an adult cell (Lamb 6LL3, better known as Dolly). In it, the nominalizations we italicize form a network that conveys a significant part of the meaning of the passage, the importance of process:

> *Development* of embryos produced by nuclear *transfer* depends upon the *maintenance* of normal ploidy and creating the conditions for developmental *regulation* of gene *expression.* These *responses* are both influenced by the cell-cycle stage of donor and recipient cells and the *interaction* between them (reviewed in ref. 9). A *comparison* of *development* of mouse and cattle embryos produced by nuclear *transfer* to oocytes or enucleated zygotes suggests that a greater *proportion* develop if the recipient is an oocyte. This may be because factors that bring about *reprogramming* of gene *expression* in a transferred nucleus are required for early *development* and are taken up by the pronuclei during *development* of the zygote.

This passage exhibits another defining characteristic of current scientific English: its nominalizations are often embedded in long and complex noun phrases, such as "development of embryos produced by nuclear transfer" and "a comparison of development of mouse and cattle embryos produced by nuclear transfer to oocytes or enucleated zygotes."

The typical complex noun phrase in scientific English consists of a head noun (*development* and *comparison* in the above examples) with multiple modifiers to the right or left, sometimes in both positions. One of the main grammatical differences in sentences in modern scientific English compared to its precursor is the frequent use of such noun phrases as the subject. Handling them adroitly is essential to good scientific writing.

We now turn to the thought process behind how an author might construct a typical sentence in scientific English—that is, two complex noun phrases connected by a fairly common main verb (*to determine*). For that purpose, we borrow the first sentence in an article by Fred Sanger and his colleagues (1977) reporting the first ever sequencing of an entire genome—the bacteriophage called ΦX_{174}, a virus that infects bacteria. The article begins with an abstract distilling the essence of the discovery. Its first sentence is as follows:

> A DNA sequence for the genome of bacteriophage ΦX_{174} of approximately 5,375 nucleotides [*noun phrase*] has been determined [*verb*] using the rapid and simple "plus and minus" method [*noun phrase*].

How was such a sentence constructed? Let's pretend the author has not yet written a word. We are looking over his shoulder at the computer terminal and spurring him on by asking questions about his research as he attempts to translate it into written words.

Our first question is "Given that this is the first sentence in the entire paper, could you choose a single word that describes either the content of the paragraph it heads or even the whole paper?" The answer he gives us is the nominalization "sequence" (not coincidentally, the nucleus noun in the article's title).

"A sequence of what?" we ask. His answer is "A DNA sequence."

That raises another question. "There are an infinite number of DNA sequences. Could you be more specific?" He amplifies, "A DNA sequence for the genome of bacteriophage ΦX_{174}."

We could ask about the definitions of the technical terms here, but we will assume our readership of molecular biochemists knows these terms already. However, even experts in the field might not know anything much about ΦX_{174} other than that it is a bacteriophage. "Could you be a little more specific?" we wonder. His answer is to elaborate with further technical detail: "A DNA sequence for the genome of bacteriophage ΦX_{174} of approximately 5,375 nucleotides." (This long and complex noun phrase, the subject of the sentence, is also, not coincidentally, the subject of the article as a whole.)

Our next question is "What's the significance of '5,375 nucleotides'"? He answers, "My readership understands that this quantity signifies a simple genome, one well suited to the task at hand."

We take him at his word but then ask, "What's the task at hand?" His response is "To determine all the 5,375 nucleotides in this genome; this would be the first time anyone has sequenced a complete genome."

"So," we wonder, "how did you determine the sequence?" He responds, "I used the plus and minus method."

"Why that method?"

"Because it is rapid and simple."

"What is this method?"

"I will provide further details later in the paper."

We have run out of questions, so we can now piece together the author's answers :

> *A DNA sequence for the genome of bacteriophage* ΦX_{174} *of approximately 5,375 nucleotides* has been determined using *the rapid and simple "plus and minus" method.*

The words in italics constitute the complex noun phrases in the sentence. The two key questions all readers want answered at the beginning of a scientific article are these: What is your major claim? How did you arrive at that claim? This typical example of scientific English answers both those questions succinctly.

Another defining feature of current scientific English is a variant of the complex noun phrase, the noun string. In standard as opposed to scientific English, writers normally qualify and expand on the meaning of a single head noun by means of modifying words added to the right. Let's start with the word "experiment." In standard English, an author might add a qualifying phrase after that noun, linked to that noun by means of prepositions and conjunctions:

> experiment *at high temperature and high pressure*

In this noun phrase, the modifier to the right of the head noun has no verb. A dependent clause can also be added to this construction:

> experiment at high temperature and high pressure *that produced* NO_2

In current scientific English, as distinguished from standard English, writers routinely transfer phrases and clauses from the right to the left of the head noun. You find

> *high-temperature high-pressure* experiment that produced NO_2

Or

> *high-temperature, high-pressure, NO_2-producing* experiment

Before the twentieth century, scientists rarely used such noun strings—nominal constructions in which prepositions, conjunctions, and pronouns disappear. On average, scientific prose has about three such constructions per hundred words.

The preference in current scientific English for nominalizations, long and complex noun phrases, and noun strings leads to the concomitant of this tendency: a preference for commonplace verbs in main clauses. Here is an example from a classic theoretical article on superconductivity by Bardeen, Cooper, and Schrieffer (1957), with the main verbs in italics:

> A theory of superconductivity *is presented,* based on the fact that the interaction between electrons resulting from virtual exchange of phonons *is* attractive when the energy difference between the electrons states involved *is* less than the phonon energy, $\hbar\omega$. It *is* favorable to *form* a superconducting phase when this attractive interaction *dominates* the repulsive screened Coulomb interaction. The normal phase *is described* by the Bloch individual-particle model. The ground state of a superconductor, *formed* from a linear combination of normal state configurations in which electrons *are* virtually *excited* in pairs of opposite spin and momentum, *is* lower in energy than the normal state by amount proportional to an average $(\hbar\omega)^2$, consistent with the isotope effect. A mutually orthogonal set of excited states in one-to-one correspondence with those of the normal phase *is obtained* by *specifying* occupation of certain Bloch states and by *using* the rest to form a linear combination of virtual pair configurations. The theory *yields* a second-order phase transition and a Meissner effect in the form *suggested* by Pippard. Calculated values of specific heats and penetration depths and their temperature variation *are* in good agreement with experiment. There *is* an energy gap for individual-particle excitations which *decreases* from about $3.5kT_c$ at $T = 0°K$ to zero at T_c. Tables of matrix elements of single-particle operators between the excited-state superconducting wave functions, useful for perturbation expansions and calculations of transition probabilities, *are given.*

The most common verb of all, *to be,* appears repeatedly. Most of the other verbs appear frequently in all scientific articles: *present, describe, form, yield, suggest, decrease, give.*

In addition to its high-frequency verbs, the passage is typical in that it is rife with complex noun phrases and noun strings, culminating with a complex noun phrase dense even by the standards of the physics literature:

"tables of matrix elements of single-particle operators between the excited-state superconducting wave functions, useful for perturbation expansions and calculations of transition probabilities."

Heavy Reliance on Technical Terms

We come now to the most visible characteristic of current scientific English, its employment of a vast array of technical terms. Without question technical terms constitute the crucial element in defining the difficulty most of us experience in reading this prose, a difficulty experienced not only by the general public but by scientists outside the special field with which a particular article deals. Here is a highly technical passage by Chien Liu and colleagues (2001) on the halting of light for the purpose of information storage:

> During the storage time, information about the amplitude of the probe field is contained in the population amplitudes defining the atomic dark states. Information about the mode vector of the probe field is contained in the relative phase between different atoms in the macroscopic sample. The use of cold atoms minimizes thermal motion and the associated smearing of the relative phase during the storage time. (We obtain storage times that are up to 50 times larger than the time it takes an atom to travel one laser wavelength. As seen from equation (1), the difference between the wavevectors of the two laser fields determines the wavelength of the periodic phase pattern imprinted on the medium, which is 10^5 times larger than the individual laser wavelengths).

The technical terms in this passage—"probe field," "population amplitudes," "mode vector," "relative phase," "laser wavelength," "wavevector," and "periodic phase pattern"—cannot be clarified merely by resorting to a dictionary. This is also true about many of its "ordinary" expressions, really technical terms in disguise: "storage time," "information," "dark state," "cold atoms," "smearing." And we have not yet reached equation 1! The passage yields its meaning only to those with an intimate acquaintance with applied quantum physics. But that is all right. The passage is meant only for those with an intimate acquaintance with quantum physics.

Conclusion and Checklist

The writing in even the best scientific periodicals routinely violates the writing guidelines routinely expounded by respected authorities. Within the typical longish sentences of scientific writing, one typically find verbs in the passive voice, nominalizations, complex noun phrases, weak verbs, noun strings, and a host of technical terms. We believe that is not because scientist-writers are needlessly obscure or incompetent communicators. Rather, the content drives the writing style.

Still, all these features do get routinely overused and abused in scientific English, making an already complex message more complex than need be. Most scientific prose does benefit from judicious trimming of these elements. Our advice is not to avoid or even minimize them at all costs but to learn how to use them effectively. As you begin to revise, then, we recommend that you weigh the following:

1. For each main verb in the passive voice, ask what verb voice works best—should the grammatical subject name the actor or the material being acted upon? Do not hesitate to use the passive voice (agentless) or name the agent—*I* or *we* or *Smith et al.*—as the situation demands. An important competing consideration involves controlling the flow of information for the purpose of closely linking one sentence with the previous—a topic we save for the next chapter.
2. For each nominalization such as *development, observation,* and *evaluation,* ask whether you ought to covert it into a verb. That revision can shorten the sentence and replace a weak verb: compare "A typical *requirement* of image *formation* is . . . " with "Image formation typically requires . . . " or "Forming an image typically requires . . . " But remember that the abstract nature of scientific writing dictates a higher density of such words than everyday writing.
3. For each complex noun phrase in a given paragraph, ask whether they are linked in a way that makes your main point clearly and concisely. If not, look to simplify the information load carried by the noun phrases. Take the earlier example of "*Tables of matrix elements of single-particle operators between the excited-state superconducting wave functions, useful for perturbation expansions and calculations of transition probabilities,* are given." Might not that make for easier reading as "We provide tables listing matrix elements of single-particle operators between the excited-state superconducting wave functions. These matrix elements are useful for perturbation expansions and calculations of transition probabilities"?
4. For each noun string, ask whether it will cause your readers any hesitation in grasping the various interrelationships. If so, revise by unpacking and making the connections more explicit. For example, "planar graphite fused six-membered ring structure" might be clearer as "the planar graphite structure consisting of a fused six-member ring," even though it is not as concise.

5. For each technical word or phrase, ask whether your intended audience really will understand its meaning. You may need to define it at first mention or choose a different wording. To theoretical physicists, the following expression is plain as day: "matrix elements of single-particle operators between the excited-state superconducting wave functions." For chemists with an interest in developing new superconducting materials, it might not be. For the rest of us, it is a foreign language.

EXERCISE

The following technical passage concerns the development of a strain of laboratory mice able to resist cancer cells:

> Despite many decades of intense research, the mechanism behind the spontaneous regression of cancer in humans and animals has remained a mystery. Zheng Cui and colleagues have bred a colony of lethal-cancer-cell-resistant BALB/c mice that exhibits spontaneous regression of advanced cancer. Mediation of this capability was found to occur by a massive infiltration of host leukocytes, but the discovery of the gene that results in the concurrence of the anticancer innate immunity has not yet been made. Remarkably, the mice-based anticancer immune system cells were injected into untreated mice, which then showed complete resistance to lethal cancer cells.

What sentences have verbs in the passive voice? What nouns are nominalizations? What words or terms would you consider "technical"? Are there complex noun phrases? Any noun strings? See if you can improve this passage by revising in accord with the above checklist.

15 Improving Scientific English

In an article titled "The Science of Scientific Writing," George Gopen and Judith Swan (1990) wrote that "complexity of thought need not lead to impenetrability of expression." We could not agree more. Yet as Gopen and Swan also no doubt recognize, the key question for scientific writing is "Impenetrable to whom?" There is no escaping the fact that current scientific English exhibits a high level of what linguists call "cognitive complexity"—that such English, while adhering to the conventions of English grammar and usage, deviates from what most nonscientists regard as standard formal English. Indeed, it deviates from what many *scientists* regard as standard formal English. Nonetheless, even with the stylistic constraints and complexities that have evolved over time, as discussed in chapter 14, contemporary scientific prose is capable of transparency to its intended audience.

We find seven basic guidelines helpful for achieving that end; accompanying each we provide a question or two to test whether the guideline has been followed. We do not claim that these are the only guidelines or questions with which to improve scientific English. We claim only that they are easily comprehended and, after a little practice, easily applied.

1. Add context. Have you provided sufficient background to enable readers to understand any new technical terms or concepts?
2. Explain the significance of your measurements and observations. For any results presented, have you provided the reader with a firm basis for evaluating their significance?
3. Be precise. Have you avoided vague generalizations such as "The temperature increased"? The critical reader will want to know by how much.
4. Specify the agent of actions if not evident from the context. Is it clear who or what performed the action?
5. Trim back excessive nominalizations and eliminate the superfluous. Can some of your nominalizations be usefully turned back into their verbs? Have you stripped away excess verbiage, including details that do not advance the paper's overall argument? Is any statement self-evident and

therefore a candidate for excision ("Curing breast cancer would relieve the suffering of millions")?

6. Weed out the ambiguous or unnecessarily complex. Have you simplified any expressions that the intended reader might find confusing? Have you used ordinary words whenever possible, instead of technical terminology?
7. Use clear transitions. Have you used appropriate words or phrases to indicate shifts in thought? Is the content of each sentence clearly linked to an earlier sentence?

To demonstrate the use of these guidelines, we will apply them to two passages: one from a first draft, another from a published article. Both passages summarize a discovery: one in the field of high-temperature superconductivity, the other in the halting and storage of light. For each passage, we give a little background information, the passage itself, and our analysis of the text. In each passage we have inserted frequent superscripts consisting of numbers and letters. In your first reading, we recommend that you ignore these. Then review the passage again in conjunction with the subsequent analysis, keyed to the superscript numbers and letters. Each number corresponds to one of our seven guidelines, each letter to a specific comment. Superscript "2c" is an example: "2" signifies the second guideline and "c" our third comment on it. After studying our analysis, read our revision and compare it with the original. (We borrowed this method of presentation from a classic writing handbook, *The Reader over Your Shoulder*, by Robert Graves and Alan Hodge.)

Draft Passage on Superconductivity

THE ORIGINAL PASSAGE

The draft text below relies heavily upon abbreviations, measurements, complex technical terms, passive verbs, and hyphenated noun phrases. The first-person pronoun *we* appears two times, but overall the writing is highly impersonal. It is also highly compressed. This passage will probably be incomprehensible to you, unless you are an expert on superconductivity. If you are such an expert, you might consider the passage readable at first glance. The sentences are short. No sentence has multiple clauses. One sentence follows logically from the next, more or less. The complexities arise for the most part because the author has assumed an intimate knowledge of his subject. Still, one might ask, could the message be made clearer? Could the author have made it more easily digested by a fellow expert, maybe even by your average materials scientist or engineer curious about the subject? That is the task before us.

We start with some background information for readers unfamiliar with high-temperature superconductivity. In the mid-1980s, scientists discovered that certain ceramic materials might be able to transport electricity without any resistance at or above 77 degrees kelvin (that is, minus 321 degrees Fahrenheit). This is called "high-temperature" superconductivity because the previous record was much lower, under 25 degrees kelvin. One of the most promising materials used in these superconductivity experiments was the compound $YBa_2Cu_3O_{7-x}$, commonly referred to as YBCO. The discovery of this ceramic caused a major stir in the scientific community. It meant that superconducting materials might be able to function with a coolant of liquid nitrogen, a relatively inexpensive refrigerant that forms at 77 kelvin and is commonly used in research labs around the world. The commercial possibilities for the electric power sector seemed very promising indeed.

After the initial elation died down, many serious technical obstacles emerged. One was how to make an electricity-carrying wire out of a ceramic that was so weak and brittle. A possible solution was to coat a metal wire with a very thin film (less than 1 micrometer) of the YBCO. But that presented another problem. This ceramic did not bond strongly to the metal. New coating methods had to be invented. In that effort, researchers found that they achieved good bonding when they inserted a specially textured ("biaxial") layer of a third material ("buffer") between the metal wire ("substrate") and the YBCO film. We now present the original draft with superscripts keyed to the analysis that follows:

> Inclined substrate deposition (ISD) offers the potential for rapid production of high-quality biaxially textured template layers suitable for YBCO-coated conductors.[1a] We have grown biaxially textured magnesium oxide (MgO) films[5a] on Hastelloy substrates by ISD[7a] at deposition rates of 20–100 Å/sec.[2a] Microscopy of the ISD-MgO films showed columnar grain structures with a roof-tile-shaped surface.[2b] A small phi-scan full-width at half maximum (FWHM)[5b] of approximately 10°[6a] was observed on MgO films.[2b] YBCO films were grown[7b] on ISD-MgO buffered substrates by pulsed layer deposition. We obtained a critical current density of $>5 \times 10^5$ A/cm^2 at 77 K on a 0.5-μm-thick YBCO film.[2c] Recently, a one-meter-long ISD-MgO metallic substrate[6b] was fabricated.[2d]

PASSAGE ANALYSIS

1. **INSUFFICIENT CONTEXT**
   a. **Inclined substrate deposition (ISD) offers the potential for rapid production of high-quality biaxially textured template layers suitable for YBCO-coated conductors.**

Is it safe to assume that readers of a specialized journal on materials science have sufficient previous knowledge to comprehend this first sentence? What is "inclined substrate deposition"? Why is "biaxial texturing" important? What is YBCO? Why should we even care about the subject? In our judgment, all readers, whatever their experience, would welcome at least a few introductory sentences to set the stage for appreciating the authors' achievement:

> *Revision*
> The superconducting material $YBa_2Cu_3O_{7-x}$ (YBCO) shows promise for use in electric power wires operating at temperatures that exceed that of liquid nitrogen (77 K). For that promise to be realized, a method is being sought to deposit biaxially textured template layers of MgO on a metal substrate suitable for later YBCO deposition. Previous researchers [1–4] have found that biaxial texturing of the MgO film strengthens the bonding of the superconducting layer.

With the new introductory sentences in place, we can now go to a slightly revised version of the original first sentence:

> *Revision*
> Smith et al. [5] report that inclined substrate deposition (ISD) offers the best route to rapid production of high-quality template layers.

Any reader uncertain about the particulars behind "inclined substrate deposition" can consult the reference indicated by the bracketed 5.

2. SIGNIFICANCE UNCLEAR

a. **. . . deposition rates of 20–100 Å/sec . . .**

How are we to assess the achievement that "20–100 Å/sec" signifies? The numbers by themselves mean nothing unless we add the explanatory phrase "acceptably rapid" before "deposition rates." The word *rapid* echoes the use of that word in the first sentence of the original.

b. **Microscopy of the ISD-MgO films showed columnar grain structures with a roof-tile-shaped surface. A small phi-scan full-width at half maximum (FWHM) of approximately 10° was observed on MgO films.**

Of these sentences we would ask: do these physical characteristics (grain structure and FWHM) indicate that the MgO films will serve their intended purpose? We also wonder about the significance of the expression "small phi-scan full-width at half maximum of approximately 10°." Is that clear?

Revision
Our analysis of these films revealed near ideal surface features for YBCO deposition: microscopy showed columnar grain structures with a roof-tile-shaped surface, and a small phi-scan yielded a full-width at half maximum of approximately 10°, confirming good biaxial texture.

c. . . . a critical current density of $>5 \times 10^5$ A/cm^2 at 77 K on a 0.5-µm-thick YBCO film.

We might assume that passing more than 500,000 amperes of current through a square centimeter is obviously impressive. But we can easily add explanatory text to make that point more clearly:

Revision
Measurements on a thin (0.5-µm) YBCO film met our goal of $>5 \times 10^5$ A/cm^2 for the critical current density at 77 K.

d. . . . a one-meter-long ISD-MgO metallic substrate was fabricated.

What is the significance of fabricating a piece "one meter long"? Does a meter length have some commercial use? Again, adding an explanatory phrase, in accord with our second guideline, helps to clarify:

Revision
As evidence of the commercial potential for ISD, we recently fabricated a wire sample of ISD-MgO metallic substrate that is one meter long.

5. SUPERFLUOUSNESS

a. . . . magnesium oxide (MgO) films . . .

We would remove the words "magnesium oxide" on the grounds that the abbreviation MgO is perfectly understandable by our technically literate audience.

b. . . . full-width at half maximum (FWHM) . . .

We would drop the abbreviation FWHM from the original as superfluous information since it does not appear again in the passage. Of course, if FWHM were to appear in a subsequent paragraph, it should remain as is. Abbreviations like this are definitely useful in scientific writing. However, when used to excess or indiscriminately, they can impede understanding by demanding too much of the readers' memory.

6. AMBIGUITY

a. . . . a small phi-scan full-width at half maximum (FWHM) of approximately 10° . . .

This expression took us awhile to unravel. Is it a small phi-scan or small full-width at half maximum? We would revise the phrase to "a small phi-scan yielded a full-width at half maximum of approximately 10°." We do assume that a knowledgeable reader knows the meaning of "phi-scan" and "full-width at half maximum."

b. . . . a one-meter-long ISD-MgO metallic substrate . . .

We can express this long noun string more clearly as "a wire sample of ISD-MgO metallic substrate that is one meter long." In this case, we would opt for clarity over conciseness.

7. POOR TRANSITIONS

a. **We have grown biaxially textured MgO films on Hastelloy substrates by ISD . . .**

We do not learn the connection to inclined substrate deposition, the subject of the first sentence, until near the end of the second sentence. We can fix that problem with an introductory phrase tying the two sentences together: "Using this technique, we have grown . . . "

b. **YBCO films were grown . . .**

This sentence introduces a new topic without any word or phrase warning us. What about "Next, we grew YBCO films on . . . "?

REVISION OF ORIGINAL

(1) The superconducting material $YBa_2Cu_3O_{7-x}$ (YBCO) shows promise for use in electric power wires operating at temperatures that exceed that of liquid nitrogen (77 K). (2) For that promise to be realized, a method is being sought to deposit biaxially textured template layers of MgO on a metal substrate suitable for later YBCO deposition. (3) Previous researchers [1–4] have found that biaxial texturing of the MgO film strengthens the bonding of the superconducting layer. (4) Smith et al. [5] report that inclined substrate deposition (ISD) offers the best route to rapid production of high-quality template layers. (5) Using this technique, we have grown biaxially textured MgO films on Hastelloy substrates at acceptably rapid deposition rates, 20–100 Å/sec. (6) Our analysis of these films revealed near ideal surface features for YBCO deposition: microscopy showed columnar grain structures with a roof-tile-shaped surface, and a small phi-scan yielded a full-width at half maximum of approximately 10°, confirming good biaxial texture. (7) Next, we grew YBCO films on ISD-MgO buffered substrates by pulsed layer deposition. (8) Measurements on a thin (0.5-μm) YBCO film met our target of $>5 \times 10^5$ A/cm² for the critical current density at 77 K.

> (9) As evidence of the commercial potential for ISD, we recently fabricated a wire sample of ISD-MgO metallic substrate that is one meter long.

Our revision reveals the logical structure of the original: logic and grammar are now in synch. In our revision, sentence 2 specifies what was needed to realize the "promise" defined in sentence 1. Sentence 3 clarifies sentence 2. Sentence 4 presents a specific method for realizing the promise mentioned in sentences 1 and 2. Sentences 5 and 6 explain what the authors accomplished by applying the fabrication method in sentence 2. Sentence 7 describes the logical next step after the success summarized in sentences 5 and 6. Sentence 8 reports a key performance measurement of the material mentioned in sentence 6. Sentence 9 mentions yet another milestone on the road to developing a commercial product. A coherent story emerges as the paragraph unfolds. It gives us a progress report on the development of a superconducting material.

Our revision is considerably longer than the original because we added context-setting information and amplified some expressions for the sake of readability. In the process of revision, we also made a few guesses about what scientist-authors and readers would be thinking. But we believe our revision achieves our goal of making the original more easily digested by those who might be interested in a progress report on the subject matter. The draft is now ready for publication. We hope you agree.

Published Passage on the Halting and Storage of Light Pulses

THE ORIGINAL PASSAGE

Now we will apply our guidelines to the first paragraph in an article that appeared in *Nature* concerning the stopping of light (Liu et al. 2001). The text itself has much to admire: not only the science but the composition as well. The passage follows the standard template of scientific introductions, the subject of chapter 1:

> Electromagnetically induced transparency [1–3] is a quantum interference effect that permits the propagation of light[5a] through an otherwise opaque atomic medium; a "coupling" laser is used to create the interference necessary to allow the transmission of resonant pulses[5b] from a "probe" laser.[1a] This technique has been used [4–6] to slow and spatially compress light pulses by seven orders of magnitude, resulting in their complete localization and containment[5c] within an atomic cloud. Here we use electromagnetically induced transparency[7a] to bring laser pulses to a complete stop in a magnetically trapped, cold cloud[3] of sodium atoms[1b] [4]. Within

> the spatially localized pulse region, the atoms are in a superposition state determined by the amplitudes and phases of the coupling and probe laser fields. Upon sudden turn-off of the coupling laser, the compressed probe pulse is effectively stopped;[5d/6a] coherent information initially contained in the laser fields is "frozen" in the atomic medium for up to 1 ms. The coupling laser is turned back on at a later time[6b] and the probe pulse is regenerated:[5e] the stored coherence is read out and transferred back into the radiation field. We present a theoretical model that[7b] reveals that the system is self-adjusting[6c] to minimize dissipative loss during the "read" and "write" operations. We anticipate applications[7c] of this phenomenon for quantum information processing.[6d]

Before beginning our analysis, which will inevitably point to violations of our guidelines, we would like emphasize that this passage is well-organized throughout and is, in general, well-written. Take the second sentence as an instance of exemplary scientific prose:

> This technique has been used [4–6] to slow and spatially compress light pulses *by seven orders of magnitude,* resulting in their complete localization and containment within an atomic cloud. (our emphasis)

The phrase "to slow and spatially compress light pulses by seven orders of magnitude" is a good example of adherence to guideline 3: be precise. If the authors had written only "to slow and spatially compress light pulses," we might have legitimately wondered by how much. Consider also the phrase "This technique has been used [4–6] to slow." We might be tempted to criticize its use of the passive voice, to ask, "Who used it, the present writers or others?" But the bracketed numbers point to references with those names: because context has provided clarification, guideline 4 has not been violated. Moreover, "this technique," placed in the subject position, forms a better link to the previous sentence than would the names of the scientists who used it (guideline 7). We would consider it ill advised to change from the passive to active voice.

When reading our passage analysis, we ask that you bear in mind three points. First, very little published prose is beyond criticism. There always seems to be a clearer, more succinct wording. Second, there are limits to the amount of time that even the most conscientious scientists can devote to fiddling with their English in an illusory quest for perfection. Third, we are not specialists in the field of quantum physics or laser technology, and our revisions may unintentionally distort the authors' intended meaning. Nevertheless, we are confident we have identified some communicative areas where there is room for improvement in the interest of clarity.

PASSAGE ANALYSIS

INSUFFICIENT CONTEXT

a. Electromagnetically induced transparency [1–3] is a quantum interference effect that permits the propagation of light through an otherwise opaque atomic medium; a "coupling" laser is used to create the interference necessary to allow the transmission of resonant pulses from a "probe" laser.

This first sentence defines the technical term "electromagnetically induced transparency." But readers do not yet have any means of telling why this term is being defined. Addition of some context-setting text might help ease us into this highly technical first sentence:

Revision

Light pulses are the fastest and most robust carriers of information, but they are difficult to localize and store. The present letter reports the demonstration of a technique that traps, stores, and releases excitations carried by light pulses. This light storage technique is based on a recently demonstrated phenomenon of ultraslow light velocity, which is made possible by electromagnetically induced transparency [1–3].

It can be argued, of course, that for an audience of applied quantum physicists—the intended audience—the original first sentence is an appropriate beginning; more is not necessary. We would merely point out that our new beginning is adapted from the introduction to an article in *Physical Review Letters* concerning a light-stopping technique employed by another group of Harvard researchers (Phillips et al. 2001).

b. a magnetically trapped, cold cloud of sodium atoms . . .

What is a "cold cloud of sodium atoms"? Why "sodium atoms"? We derived an expansion of this phrase from an article in *Scientific American* by Lene Vestegaard Hau (2001), the lead researcher of the team that produced the article.

Revision

. . . a magnetically trapped cloud of sodium atoms cooled to 500 billionths of a degree above absolute zero [4], a temperature at which a Bose-Einstein condensate forms, one in which all of the atoms behave synchronously.

3. IMPRECISION

. . . cold cloud . . .

How cold is "cold"? See comment 1b for the answer.

5. SUPERFLUOUSNESS AND EXCESSIVE NOMINALIZATION

a. that permits the propagation of light . . .

Here we have a nominalization ("propagation") that we can convert to a verb for conciseness: "that permits light to propagate . . . " See chapter 14.

b. to allow the transmission of resonant pulses . . .

Another nominalization from the same sentence can go: "to transmit resonant pulses . . . "

c. their complete localization and containment . . .

We can delete "and containment" as superfluous because "complete localization" implies "containment," though perhaps there is a subtle point we are missing.

d. Upon sudden turn-off of the coupling laser, the compressed probe pulse is effectively stopped . . .

Wouldn't it be simpler to say: "Turning off the laser halts the compressed light"? Does that alter the intended meaning?

e. The coupling laser is turned back on at a later time and the probe pulse is regenerated . . .

To maintain parallelism in sentence structure to our suggested revision in comment 5d above, we would change to "Turning the coupling laser back on . . . "

6. AMBIGUITY AND VAGUENESS

a. probe pulse is effectively stopped . . .

We would eliminate the word *effectively* because it means both "in an effective manner" and "for all practical purposes," that is, *not* completely.

b. on at a later time . . .

We would question what a "later time" means. Does it mean during or after the 1 millisecond of capture mentioned in the previous sentence? We decided on the former, though our choice could easily be mistaken.

c. the system is self-adjusting . . .

What "system"? We would recommend the more specific "experimental system."

d. . . . this phenomenon for quantum information processing.

What "phenomenon"? Is it the "self-adjusting" of the previous sentence or all the phenomena of the stopping and storing of light just described? We suspect the latter.

7. POOR TRANSITIONS

a. **Here we use electromagnetically induced transparency . . .**

A smoother transition would be "With this technique, we have brought . . . "

b. **We present a theoretical model that reveals that . . .**

The subject has changed from the authors' experimental work to a theoretical model, but what's important here, we suspect, is not the theoretical model per se but the finding that the system self-adjusts. A better transition might be "According to our theoretical model, the experimental system self-adjusts . . . "

c. **We anticipate applications . . .**

The subject has again changed, this time to applications of the research. We could better signal that change by beginning this sentence "Our research findings . . . "

REVISION OF ORIGINAL

> (1) Light pulses are the fastest and most robust carriers of information, but they are difficult to localize and store. (2) The present letter reports the demonstration of a technique that traps, stores, and releases excitations carried by light pulses. (3) This light storage technique is based on a recently demonstrated phenomenon of ultraslow light velocity, which is made possible by electromagnetically induced transparency [1–3]. (4) Electromagnetically induced transparency is a quantum interference effect that permits light to propagate through an otherwise opaque atomic medium: a "coupling" laser is used to create the interference necessary to transmit resonant pulses from a "probe" laser. (5) This technique has been used [4–6] to slow and spatially compress light pulses by seven orders of magnitude, resulting in their complete localization within an atomic cloud. (6) By using this technique, we have brought laser pulses to a complete stop in a magnetically trapped cloud of sodium atoms cooled to 500 billionths of a degree above absolute zero [4], a temperature at which a Bose-Einstein condensate forms, a medium in which all of the atoms behave synchronously. (7) Within the spatially localized pulse region, the atoms are in a superposition state determined by the amplitudes and phases of the coupling and probe laser fields. (8) Turning off the coupling laser halts the compressed probe pulse; coherent information initially contained in

the laser fields is "frozen" in the atomic medium for up to 1 ms. (9) Turning the coupling laser back on within this time regenerates the probe pulse; the stored coherence is read out and transferred back into the radiation field. (10) According to our theoretical model, the experimental system self-adjusts to minimize dissipative loss during the "read" and "write" operations. (11) Our research findings have important potential applications in quantum information processing.

Our revision reveals the logical structure of the original. Sentence 1 establishes a physics problem. Sentence 2 states in brief the authors' solution to the problem in sentence 1. Sentence 3 defines the "light storage technique" behind the solution. Sentence 4 defines the technical term introduced at the end of sentence 3. Sentence 5 presents what earlier researchers achieved with the technique in sentence 4. Sentence 6 presents what the authors did that topped the earlier research in sentence 5. Sentences 7–10 elaborate on sentence 6 by explaining what the authors achieved in the lab and by theoretical analysis. Sentence 11 alludes to the benefit to society of the achievement defined in sentences 6–10.

As with our superconductivity example, our revision is longer than the original. In both cases, their authors could argue, plausibly, that we should have left well enough alone. Further, we were not constrained, as the authors of the halting-of-light article were, by the parsimonious word count dictated by *Nature* for article length or by the time pressure to establish priority speedily for an important discovery. Our purpose here is not to second-guess but to provide an example of how applying our guidelines to a passage can reduce the mental processing required by the reader and widen the article's audience.

Conclusion

The goal of scientific English is to achieve clarity. To do so is not to fight against the agreed-upon constrains of current scientific English: its heavy reliance on technical terms and its systematic employment of the passive voice, nominalization, complex noun phrases, noun strings, and weak verbs. Rather, the task is to achieve clarity within these constraints. In this chapter we offer guidelines for achieving clarity when so constrained. We do so to demonstrate that this achievement is within everyone's reach. We are not asserting that ours is more than a beginning; we believe that even greater clarity is within everyone's reach, though not without considerable effort. But for that another book of instruction is necessary. Fortunately, in our need we can turn to that masterpiece of pedagogy, Joseph Williams's *Style*.

EXERCISE

Here is a concocted superconductivity paragraph with each sentence numbered for easy reference and discussion.

> (1) In the mid-1980s, novel ceramic materials that transport electricity without loss ("superconducting") at a much higher temperature than ever achieved before were discovered. (2) Liquid nitrogen forms at 77 kelvin. (3) $YBa_2Cu_3O_{7-x}$, commonly referred to as YBCO, is a superconducting material. (4) Since then, scientists have created YBCO compositions with even higher superconducting temperatures. (5) They suggest that superconducting materials could function with a coolant of liquid nitrogen, a relatively inexpensive refrigerant. (6) A major stir in the scientific community arose because of these discoveries.

Are you able to easily follow the flow of information? Are the transitions from one idea to the next clear? They should not be, because we purposely wrote the paragraph to confuse. Go through it and check the logical connections between each sentence and its preceding sentence.

Answer

Here is our analysis of the missed connections:

Sentence 2. What does liquid nitrogen have to do with the superconducting materials in sentence 1?
Sentence 3. What is connection between YBCO and 77 kelvin in sentence 2?
Sentence 4. How is "then" connected to YBCO in sentence 3, if at all? Temperatures "higher" than what?
Sentence 5. Does "they" refer to scientists in sentence 4? If so, it makes no sense.
Sentence 6. What "discoveries"? Only one "discovery" stated in sentence 5?

CHECKLIST

Now, with our checklist below in mind, see if you can revise the sentences to improve the continuity of thought.

- Add context to provide sufficient background so that your readers will understand any new technical terms or concepts.
- Explain the significance of data reported.
- Quantify how much or how many for comparative expressions (*hot, cold, high, low, increase, decrease,* etc).
- Specify the agent of actions if not evident from the context.

- Trim back excessive nominalizations and eliminate the superfluous.
- Revise to root out the ambiguous or unnecessarily complex.
- Check for logical connections within and between sentences.

A Final Thought

In closing, we wish to return briefly to the question with which this book began: Why bother? We offer one more reason: there is a tradition to uphold. Henry Oldenburg, first editor of the first scientific periodical in English, commented on the importance of communication to science as long ago as 1665: "[T]here is nothing more necessary for promoting the improvement of Philosophical [Scientific] Matters, than the communicating to such, as apply their Studies and Endeavours that way." Within the enormous body of literature created by Oldenburg and countless others, you can find the name of every researcher of importance: Isaac Newton, Robert Boyle, Antoine Lavoisier, Charles Darwin, Louis Pasteur, Marie Curie, Albert Einstein, Werner Heisenberg, Linus Pauling, Rosalind Franklin, Richard Feynman, and Francis Crick, to name an illustrious dozen. You may never scale the same heights scientifically. Few ever do. But you can seek to approach them in terms of clarity and forcefulness of argument. Our hope is that this book helps you do that.

ACKNOWLEDGMENTS

We both have benefited greatly from the books on writing by the late Joseph M. Williams (notably *Style* and *The Craft of Research*). We have also had the pleasure of his company on several occasions. Moreover, Alan Gross participated in an NEH Summer Seminar with Prof. Williams, for him a career-altering experience. What struck us the most about Williams's personality was his insatiable curiosity and uncanny capacity to see below the surface of things. His genius, we believe, lay in his ability to question conventional wisdom on communicative practices. We hope a little of that critical intelligence rubbed off on us.

Both authors wish to thank the several dozen scientists who graciously shared their PowerPoint presentations and research proposals with us. We learned something from all of them, although we could not use them all. Especially helpful in their response to our queries were Allan I. Basbaum, University of California, San Francisco; Adrian Bejan, Duke University; Terrence J. Collins, Carnegie-Mellon University; Daniel Dzurisin, U.S. Geological Survey; Fred Gould, North Carolina State University; Kip Hodges, Arizona State University; Roald Hoffmann, Cornell University; Terry L. Hunt, University of Hawai'i–Manoa; Richard M. Iverson, U.S. Geological Survey; David Jewitt, Institute for Astronomy, University of Hawaii; Abraham Loeb, Harvard University; Shane Ross, Virginia Tech; Thomas D. Seeley, Cornell University; Solomon H. Snyder, Johns Hopkins University; and Michael Wolfe, Harvard University.

One prominent scientist who did not send us a proposal did offer an intriguing comment via e-mail: "I think . . . you will have a hard time finding anyone who can successfully write a proposal that is likely to be funded. It has become a random process." We can understand his deep skepticism: nowadays highly worthwhile proposals get routinely rejected. Nonetheless, we are still naive enough to believe that success is more likely for those able to construct a credible and clearly written argument, one capable of persuading others that the funding of their research will benefit science and society.

Joseph Harmon wishes to acknowledge the scientific and support staff at Argonne National Laboratory, in particular those present during his many years working there for the Chemical Engineering and Energy Technology divisions. From these scientists he learned a tremendous amount about how experts can best communicate their specialized knowledge to other experts, experts in other disciplines, students, and the public. The reading and writing necessary to produce this book, however, were performed independently of Argonne and the U.S. Department of Energy.

Lastly, we want to express our deep gratitude to executive editor Christie Henry, University of Chicago Press, and the reviewers she enlisted to critique our manuscript. Also to be thanked are Aydin Mohseni for readying the illustrations for publication and Ruth Goring for copy editing. The usual disclaimer applies.

REFERENCES

Agnor, Craig, and Douglas Hamilton. 2006. Neptune's capture of its moon Triton in a binary-planet gravitational encounter. *Nature* 441:192–94.

Avery, Oswald T., Colin M. MacLeod, and Maclyn McCarty. 1944. Studies on the chemical nature of the substance inducing transformation of pneumococcal types. *Journal of Experimental Medicine* 79:137–58.

Baade, W., and F. Zwicky. 1934. Super-novae and cosmic rays. *Physical Review* 45:138.

Bach, Peter B., James R. Jett, Ugo Pastorino, Melvyn S. Tockman, Stephen J. Swensen, and Colin B. Begg. 2007. Computed tomography screening and lung cancer outcomes. *Journal of the American Medical Association* 297:953–61.

Baltimore, David. 1970. Viral RNA-dependent DNA polymerase: RNA-dependent DNA polymerase in virions of RNA tumour viruses. *Nature* 226:1209–11.

Bandura, Albert, Dorothea Ross, and Sheila A. Ross. 1961. Transmission of aggression through imitation of aggressive models. *Journal of Abnormal and Social Psychology* 63:575–83.

———. 1963. Imitation of film-mediated aggressive models. *Journal of Abnormal and Social Psychology* 66:3–11.

Bardeen, J., L. N. Cooper, and J. R. Schrieffer. 1957. Theory of superconductivity. *Physical Review* 108:1175–204.

Basbaum, Allan I. 2000. Spinal cord nociceptive processing in mice. Proposal for National Institutes of Health.

Basbaum, Allan I., and David Julius. 2006. Toward better pain control. *Scientific American* 294:60–67.

Bazerman, Charles. 1988. *Shaping written knowledge: The genre and activity of the experimental article in science.* Madison: University of Wisconsin.

Bejan, Adrian. 2006. Proof-of-concept: New 4th year undergraduate course on "Constructal Design of Energy-System Configuration." Proposal for National Science Foundation.

Bligh, E. G., and W. J. Dyer. 1959. A rapid method of total lipid extraction and purification. *Canadian Journal of Biochemistry and Physiology* 8:911–17.

Boag, P. T., and P. R. Grant. 1981. Intense natural selection in a population of Darwin's finches (Geospizinae) in the Galápagos. *Science* 214:82–85.

Booth, Wayne C., Gregory G. Colomb, and Joseph M. Williams. 1995. *The craft of research.* Chicago: University of Chicago Press.

Boyle, Robert. 1664. *Experiments and Considerations concerning Colours.* London: Henry Herringman.

———. 1670. New pneumatical observations about respiration. *Philosophical Transactions* 4:2011–56.

Cacace, F., Giulia de Petris, Federico Pepi, Ivan Rossi, and Alessandro Venturini. 1996. The gas-phase reaction of nitronium ion with ethylene: Experimental and theoretical study. *Journal of American Chemical Society* 118:12719–23.

Child, Julia. 2000. *Julia's kitchen wisdom.* New York: Alfred A. Knopf.

Chu, C. W., P. H. Hor, R. L. Meng, L. Gao, Z. J. Huang, and Y. Q. Wang. 1987. Evidence for superconductivity above 40 K in the La-Ba-Cu-O compound system. *Physical Review Letters* 58:405–7.

Cleveland, William S. 1985. *The elements of graphing data.* Monterey, CA: Wadsworth.

Collins, Harry. 2004. *Gravity's shadow: The search for gravitational waves.* Chicago: University of Chicago Press.

Collins, Terence J. 2004. Understanding mechanisms of green oxidation catalysis by iron-TAML peroxide activators. Proposal for Environmental Protection Agency.

Collins, Terence J., and Chip Walter. 2006. Little green molecules. *Scientific American* 294:82–90.

Conley, Charles C. 1968. Low energy transit orbits in the restricted three-body problem. *SIAM Journal of Applied Mathematics* 16:732–46.

Curie, Pierre, Marie Sklodowska Curie, and G. Bémont, 1898. On a new, strongly radioactive substance, contained in pitchblende. *Comptes Rendus* 127:1215–17.

Daeschler, Edward B., Neil H. Shubin, and Farish A. Jenkins. 2006. A Devonian tetrapod-like fish and the evolution of the tetrapod body plan. *Nature* 440:757–63.

Darwin, Charles. 1859. *On the origin of species by means of natural selection, or the preservation of favoured races in the struggle for life.* London: John Murray.

d'Errico, Francesco, Christopher Henshilwood, Marian Vanhaeren, and Karen van Niekerk. 2005. *Nassarius kraussianus* shell beads from Blombos Cave: Evidence for symbolic behaviour in the Middle Stone Age. *Journal of Human Evolution* 48:3–24.

Doumont, Jean-luc. 2005. The cognitive side of PowerPoint: Slides are not evil. *Technical Communication* 52:64–70.

Dreifus, Claudia. 2006. Solving a mystery of life, then tracking a real life problem. *New York Times,* July 4, 2006.

Dzurisin, Daniel. 2006. The ongoing, mind-blowing eruption of Mount St. Helens. PowerPoint presentation at meeting of the Geological Society of Oregon County, Portland State University, November 17.

Einstein, A. 1905a/1998. Does the inertia of a body depend on its energy content? In *Einstein's miraculous year: Five papers that changed the face of physics,* ed. John Stachel, 161–64. Princeton: Princeton University Press.

———. 1905b/1998. On the electrodynamics of moving bodies. In *Einstein's miraculous year: Five papers that changed the face of physics,* ed. John Stachel. Princeton, NJ: Princeton University Press.

———. 1935. Can quantum-mechanical description really be considered complete? *Physical Review* 47:777–80.

Eliot, T. S. 1942/1970. Four quartets. In *Collected poems: 1909–1962.* New York: Harcourt, Brace and World.

Farman, J. C., B. G. Gardiner, and J. D. Shanklin. 1985. Large losses of total ozone in Antarctica reveal seasonal ClO_x/NO_x interaction. *Nature* 315:207–10.

Feynman, Richard. 1965. The development of the space-time view of quantum mechanics. Nobel lecture, http://nobelprize.org/nobel_prizes/physics/laureates/1965/feynman-lecture.html.

Franck, Georg. 2002. The scientific economy of attention: A novel approach to collective rationality of science. *Scientometrics* 55:3–26.

Freud, Sigmund. 1911/1913. *The interpretation of dreams.* 3rd ed. Translated by A. A. Brill. New York: Macmillan.

Galison, Peter L. 2006. Events of 1905. *Scientific American* 294:264.

Garwin, Laura, and Tim Lincoln, eds. 2003. *A century of nature: Twenty-one discoveries that changed science and the world.* Chicago: University of Chicago Press.

Gende, Scott M., and Thomas P. Quinn. 2006. The fish and the forest. *Scientific American* 294:84–89.

Gombrich, Richard F. 1989. The death of English: Letter to the editor. *New York Review of Books,* March 16, 1989.

Gopen, George D., and Judith A. Swan, 1990. The science of scientific writing. *American Scientist* 78:550–58.

Gould, Fred L. 2003. Population genetics of transgenes in insect vectors. Proposal for National Institutes of Health.

Gould, Fred, Krisztian Magori, and Yunxin Huang. 2006. Genetic strategies for controlling mosquito-borne diseases. *American Scientist* 94:238–46.

Gott, J. Richard III, Mario Jurić, David Schlegel, Fiona Hoyle, Michael Vogeley, Max Tegmark, Neta Bahcall, and Jon Brinkmann. 2003. A map of the universe. Lawrence Berkeley National Laboratory Paper LBNL-57535.

Graves, Robert, and Alan Hodge. 1979. *The reader over your shoulder: A handbook for writers of English prose.* New York: Random House.

Green, Richard E., Johannes Krause, Susan E. Ptak, Adrian W. Briggs, Michael T. Ronan, Jan F. Simons, Lei Du, Michael Egholm, Jonathan M. Rothberg, Maja Paunovic, and Svante Pääbo. 2006. Analysis of one million base pairs of Neanderthal DNA. *Nature* 444:330–36.

Gross, Alan G., Joseph E. Harmon, and Michael Reidy. 2002. *Communicating science: The scientific article from the 17th century to the present.* New York: Oxford University Press.

Harmon, Joseph E., and Alan G. Gross. 2007. *The scientific literature: A guided tour.* Chicago: University of Chicago Press.

Hau, Lene Vestergaard. 2001. Frozen light. *Scientific American* 285, no. 1: 66–73.

Heisenberg, Werner. 1927/1983. On the perceptual content of quantum theoretical kinematics and mechanics. Translated by J. A. Wheeler and W. H. Zurek. In *Quantum theory and measurement,* 62–84. Princeton, NJ: Princeton University Press.

Henschke, C. I., D. F. Yankelevitz, D. M. Libby, M. W. Pasmantier, J. P. Smith, and O. S. Miettinen. 2006. Survival of patients with stage I lung cancer detected on CT screening. *New England Journal of Medicine* 355:763–71.

Hewish, A., S. J. Bell, J. D. H. Pilkington, P. F. Scott, and R. A. Collins. 1968. Observation of a rapidly pulsating radio source. *Nature* 217:709–13.

Hodges, Kip, and Kelin Whipple. 2003. Uplift history of the Cordillera Occidental, southern Peru, from canyon geomorphology. Proposal for National Science Foundation.

Holman, Gordon. 2006. The mysterious origins of solar flares. *Scientific American* 294:38–45.

Hubble, Edwin. 1929. A relation between distance and radial velocity among extra-galactic nebulae. *Proceedings of the National Academy of Sciences* 15:168–73.

International Human Genome Sequencing Consortium. 2001. Initial sequencing and analysis of the human genome. *Nature* 409:860–921.

Iverson, Richard M. 2006. A dynamic model of seismogenic volcanic extrusion, Mount St. Helens, 2004–2005. PowerPoint presentation at fall meeting, American Geophysical Society, December 5–9.

Iverson, Richard M., Daniel Dzurisin, Cynthia A. Gardner, Terrence M. Gerlach, Richard G. LaHusen, Michael Lisowski, Jon J. Major, Stephen D. Malone, James A. Messerich, Seth C.

Moran, John S. Pallister, Anthony I. Qamar, Steven P. Schilling, and James W. Vallance. 2006. Dynamics of seismogenic volcanic extrusion at Mount St Helens in 2004–05. *Nature* 444:439–43.

Jacob, F., and J. Monod. 1961. Genetic regulatory mechanisms in the synthesis of proteins. *Journal of Molecular Biology* 3:318–56.

Jewitt, David, Scott S. Sheppard, and Jan Kleyna. 2006. The strangest satellites in the solar system. *Scientific American* 295:40–47.

Kendrew, J. C., G. Bodo, H. M. Dintzis, R. G. Parrish, H. Wyckoff, and D. C. Phillips. 1958. A three-dimensional model of the myoglobin molecule obtained by x-ray analysis. *Nature* 181:662–66.

Kevles, Daniel J. 1998. *The Baltimore case: A trial of politics, science, and character.* New York: W. W. Norton.

Kimberly, W. Taylor, Matthew J. LaVoie, Beth L. Ostaszewski, Wenjuan Ye, Michael S. Wolfe, and Dennis J. Selkoe. 2003. γ-Secretase is a membrane protein complex comprised of presenilin, nicastrin, aph-1, and pen-2. *Proceedings of National Academy of Sciences* 100:6382–87.

Kress, Gunther R., and Theo van Leeuwen. 1996. *Reading images: The grammar of visual design.* New York: Routledge.

Kroto, H. W., J. R. Heath, S. C, O'Brien, R. F. Curl, and R. E. Smalley. 1985. C_{60}: Buckminsterfullerene. *Nature* 318:162–63.

Lauterbur, P. C. 1973. Image formation by induced local interactions: Examples employing nuclear magnetic resonance. *Nature* 242:190–91.

Lévi-Strauss, Claude. 1955/1974. *Tristes tropiques.* Translated by John Weightman and Doreen Weightman. New York: Atheneum.

Liu, Chien, Zachary Dutton, Cyrus H. Behroozi, and Lene Vestergaard Hau. 2001. Observation of coherent optical information storage in an atomic medium using halted light pulses. *Nature* 409:490–93.

Loeb, Abraham, and Rennab Barkana. 2001. Observable signatures of reionization. Proposal for National Aeronautics and Space Administration.

Logothetis, Nikos K. 2008. What we can do and what we cannot do with fMRI. *Nature* 453:869–78.

Marsden, Jerrold E., and Shane D. Ross. 2006. New methods in celestial mechanics and mission design. *Bulletin of the American Mathematical Society* 43:43–73.

Mayor, Michel, and Didier Queloz. 1995. A Jupiter-mass companion to a solar-type star. *Nature* 378:355–59.

Meitner, Lise, and Otto Frisch. 1939. Disintegration of uranium by neutrons: A new type of nuclear reaction. *Nature* 143:239–40.

Montgomery, Scott L. 1996. *The scientific voice.* New York: Guilford Press.

———. 2003. *Chicago guide to communicating science.* Chicago: University of Chicago Press.

Mulshine, James L. 2008. Commentary: Lung cancer screening—progress or peril. *Oncologist* 13:435–38.

Neher, Erwin, and Bert Sakmann. 1976. Single-channel currents recorded from membrane of denervated frog muscle fibres. *Nature* 260:799–802.

Newton, Isaac. 1672. New theory about light and colors. *Philosophical Transactions* 6: 3075–87.

Nüsslein-Volhard, Christiane, and Eric Wieschaus. 1980. Mutations affecting segment number and polarity in *Drosophila. Nature* 287:795–801.

Oldenburg, Henry. 1665. The introduction. *Philosophical Transactions* 1:1–2.

Pauling, Linus, Robert B. Corey, and H. R. Branson. 1951. The structure of proteins: Two hydrogen-bonded helical configurations of the polypeptide chain. *Proceedings of the National Academy of Sciences* 37:205–11.

Peebles, P. J. E., and Joseph Silk. 1990. A cosmic book of phenomena. *Nature* 346:233–39.

Phillips, D. F., A. Fleischhaueer, A. Mair, R. L. Wadsworth, and M. D. Lukin. 2001. Storage of light in atomic vapor. *Physical Review Letters* 86:783–86.

Quinn, Thomas P., Andrew P. Hendry, and Gregory B. Buck. 2001. Balancing natural and sexual selection in sockeye salmon: Interactions between body size, reproductive opportunity and vulnerability to predation by bears. *Evolutionary Ecology Research* 3:917–37.

Robinson, Wade, Roger Boisjoly, David Hoeker, and Stefan Young. 2002. Representations and misrepresentations: Tufte and the Morton Thiokol engineers on the *Challenger. Science and Engineering Ethics* 8:59–81.

Ross, Shane D. 2003. Multiscale dynamics and phase space transport in non-integrable dynamical systems. Proposal for National Science Foundation.

———. 2006. The interplanetary transport network. *American Scientist* 94:230–37.

Sanger, F., G. M. Air, B. G. Barrell, N. L. Brown, A. R. Coulson, J. C. Fiddes, C. A. Hutchison, P. M. Slocombe, and M. Smith. 1977. Nucleotide sequence of bacteriophage fX_{174} DNA. *Nature* 265:687–95.

Seeley, Thomas D. 2001. Group decision-making in swarms of bees. Proposal for National Science Foundation.

———. 2005. House hunting by honey bees: A study in group decision-making. PowerPoint presentation at Cornell University, Department of Entomology, April 1.

Seeley, Thomas D., P. Kirk Visscher, and Kevin M. Passino. 2006. Group decision making in honey bee swarms. *American Scientist* 94:220–29.

Shubin, Neil H., Edward B. Daeschler, and Farish A. Jenkins. 2006. The pectoral fin of *Tiktaalik roseae* and the origin of the tetrapod limb. *Nature* 440:764–71.

Siegel, Warren. 1986. The super G-string. In *Workshop on unified string theories,* 729–37. Singapore: World Scientific Publishing.

Spingel, Volker, Simon D. M. White, Adrian Jenkins, Carlos S. Frenk, Naoki Yoshida, Liang Gao, Julio Navarro, Robert Thacker, Darren Croton, John Helly, John A. Peacock, Shaun Cole, Peter Thomas, Hugh Couchman, August Evrard, Jörg Colberg, and Frazer Pearce. 2005. Simulations of the formation, evolution, and clustering of galaxies and quasars. *Nature* 435:629–36.

Stafford, Ned. 2006. Scientists counter Wilmut criticisms. *Scientist,* March 2006.

Stojanovic, Milan N., and Darko Stefanovic. 2003. A deoxyribozyme-based molecular automaton. *Nature Biotechnology* 21:1069–74.

Swales, John M. 1990. Research articles in English. In *Genre analysis: English in academic and research settings,* 137–66. Cambridge: Cambridge University Press.

Taylor, J. Herbert. 1960. Nucleic acid synthesis in relation to the cell division cycle. *Annals of the New York Academy of Sciences* 90:409–21.

Thomson, James A., Joseph Itskovitz-Eldor, Sander S. Shapiro, Michelle A. Waknitz, Jennifer J. Swiergiel, Vivienne S. Marshall, and Jeffrey M. Jones. 1998. Embryonic stem cell lines derived from human blastocysts. *Science* 282:1145–47.

Tufte, Edward R. 1983. *The visual display of quantitative information.* Cheshire, CT: Graphics Press.

———. 1997. The decision to launch the space shuttle *Challenger.* In *Visual explanations,* 38–53. Cheshire, CT: Graphics Press.

——. 2003. *The cognitive style of PowerPoint: Pitching out corrupts within.* Cheshire, CT: Graphics Press.

VanLaningham, Jody, David R. Johnson, and Paul Amato. 2001. Marital happiness, marital duration, and the U-shaped curve: Evidence from a five-wave panel study. *Social Forces* 79:1313–42.

Ward, A. W. 1882. "Craft." In *Oxford English Dictionary,* 2:1128. Oxford: Oxford University Press.

Waksman, B. H., 1980. Information overload in immunology: Possible solutions to the problem of excessive publication. *Journal of Immunology* 124:1009.

Watson, J. D. 1968/1997. *The double helix: A personal account of the discovery of the structure of DNA.* London: Weidenfeld and Nicolson.

Watson, J. D., and F. H. C. Crick. 1953. Molecular structure of nucleic acids: A structure for deoxyribose nucleic acid. *Nature* 171:737–38.

Weaver, David, Moema H. Reis, Christopher Albanese, Frank Constantini, David Baltimore, and Thereza Imanishi-Kari. 1986. Altered repertoire of endogenous immunoglobulin gene expression in transgenic mice containing a rearranged mu heavy chain gene. *Cell* 45:247–59.

Wegener, Alfred. 1912. The origin of the continents. *Petermanns Geographische Mittheilungen aus Justus Perthes' Geographischer Anstalt* 58:185–95, 253–56, 305–9.

Welch, H. Gilbert, Steven Woloshin, and Lisa M. Schwartz. 2007. How two studies on cancer screening led to two results. *New York Times,* http://www.nytimes.com/2007/ 03/13/health/13lung.html.

Williams, Joseph M. 1990. *Style: Toward clarity and grace.* Chicago: University of Chicago Press.

Wilmut, I., A. E. Schnieke, J. McWhir, A. J. Kind, and K. H. S. Campbell. 1997. Viable offspring derived from fetal and adult mammalian cells. *Nature* 385:810–13.

Wilson, J. Tuzo. 1966. Did the Atlantic close and then re-open? *Nature* 211:676–81.

Woese, Carl R., Otto Kandler, and Mark L. Wheelis. 1990. Towards a natural system of organisms: Proposal for the domains Archaea, Bacteria, and Eucarya. *Proceedings of the National Academy of Sciences* 87:4576–79.

Wolfe, Michael S. 2000. Role of γ-secretase in the pathogenesis of Alzheimer's disease. Proposal for National Institutes of Health.

INDEX